A Terrible Beauty is Born

Science Spectra Book Series

Series Editor: Vivian Moses, King's College, University of London, UK

A Terrible Beauty is Born

Clones, Genes and the
Future of Mankind

Brendan Curran

Taylor & Francis
Taylor & Francis Group

LONDON AND NEW YORK

First published 2003
by Taylor & Francis
11 New Fetter Lane, London EC4P 4EE

Simultaneously published in the USA and Canada
by Taylor & Francis Inc,
29 West 35th Street, New York, NY 10001

Taylor & Francis is an imprint of the Taylor & Francis Group

© 2003 Taylor & Francis

Typeset in Optima by Wearset Ltd, Boldon, Tyne and Wear
Printed and bound in Malta by Gutenberg Press Ltd

Every effort has been made to ensure that the advice and information in this book is true and accurate at the time of going to press. However, neither the publisher nor the authors can accept any legal responsibility or liability for any errors or omissions that may be made. In the case of drug administation, any medical procedure or the use of technical equipment mentioned within this book, you are strongly advised to consult the manufacturer's guidelines.

British Library Cataloguing in Publication Data
A catalogue record for this book is available from the British Library

Library of Congress Cataloging in Publication Data
Curran, Brendan.
   A terrible beauty is born : clones, genes and the future of mankind / Brendan Curran.
      p. ; cm. – (Science spectra book series ; v. 3)
   Includes bibliographical references and index.

1. Cloning–Popular works. 2. Cloning–Moral and ethical aspects–Popular works.
[DNLM: 1. Bioethical Issues–Popular Works. 2. Cloning Organism–Popular Works. 3. Genetic Engineering–Popular Works. QH442.2 C976t 2003] I. Title.
II. Science spectra ; v. 3.

QH442.2 .C87 2003
303.48'3–dc21

ISBN 0-415-28708-1 (hbk)
ISBN 0-415-28709-X (pbk)

To Virginia and the boys
(Thank you for your love)

In memory of Mary Curran
(Recently deceased yet still a source of inspiration)

'All changed, changed utterly
A terrible beauty is born'

Easter 1916 – W. B. Yeats

Contents

Preface

She was born on 27 July 1997. Physically indistinguishable from the many thousands of other lambs born in the Scottish Highlands that year, Dolly nevertheless within months had made headline news throughout the world. Ironically, her claim to fame was that far from being unique, she was an exact replica of a sheep that had died 4 years earlier! A breast tissue sample, taken and frozen before the animal had died, had been used to produce individual cells in a test tube. Scientists had taken the nucleus containing all of the genetic information from one of these cells, injected it into an unfertilised sheep egg, implanted the new cell into a surrogate mother and succeeded in producing an animal genetically identical to the one which died 4 years earlier. The process is called nuclear transplantation, the product: an animal clone.

The media frenzy greeting the birth of Dolly was highly reminiscent of scenes played out 25 years earlier when scientists in America announced that they had managed to remove a small portion of genetic information from one organism and splice it successfully into another. That process they called *recombinant DNA technology*, the product: a *gene clone*. (What DNA is, how it works and why we use that abbreviation will all emerge in Chapter 3).

Genetic information is normally transferred 'vertically' from an organism to its offspring when it reproduces, or from one cell to the next when it divides. These new technologies have circumvented this process because together they allow some or all of the genetic information to be moved 'horizontally' – from one cell to another. That is extremely useful because it allows a large population of organisms to acquire new genetic traits within *one* generation rather than having to wait the many generations normally required to achieve a result via the vertical process. Conversely, in such manipulation experiments it is not easy to reverse anything which might go wrong; once genetic information gets into a cell it becomes hard to remove it, so genetically modified organisms give rise to offspring carrying the same modifications.

The world entered new and uncharted territory in the 1970s with the first successful demonstration of recombinant DNA technology by Herbert Boyer and Stanley Cohen, working at Stanford University. At that time, many observers (scientists and non-scientists alike) felt that such experiments should be banned because the new technology involved so many unknown parameters. What if a toxin gene were cloned by accident into a bacterium that escaped from the laboratory? Might a laboratory accident result in an uncontrollable epidemic of a cancer-causing micro-organism? Others argued that the risk of these things happening was very small and that the experiments should proceed with due care. Faced with such uncertainty a major conference was convened and a voluntary code of practice agreed.

Today, more than two decades later, we have learnt a lot more about genes and their roles in living organisms. Recombinant DNA technology has facilitated the transfer of genetic material between all sorts of organisms for a wide variety of reasons. As the sophistication of the procedures has increased, so has the impact on humans and human society of recombinant DNA technology intensified. Many human genes have been spliced into other organisms to produce vital proteins for use in alleviating human diseases; examples, including *insulin* (for the treatment of diabetes) and *human growth hormone* (for dealing with growth deficiency), are produced in bacteria while blood clotting factors to treat haemophiliacs are made in cows milk. Animals have been genetically engineered to produce organs for transplant into humans. Some of the plants that we eat or use as a source of raw materials have also been manipulated. Bacterial genes have been cloned into plants to make them resistant to pests and herbicides; plant genes have been re-regulated to increase the shelf-life of agricultural produce.

Despite the many millions of recombinant DNA experiments already carried out, the occurence of even *one* infection caused by a recombinant organism has yet to be recorded. Today the safety concerns, while still germane to regulatory considerations, are gradually being replaced by ethical ones: who should have access to an individual's genetic information? Who decides what a dysfunctional gene is? Should gene therapy be applied to germ line cells? The list goes on. . .

The successful production of Dolly has added to these worries. Scientists wish to use this technology to gain valuable insights into how cells are regulated and at the same time develop it for use in animal breeding both for agricultural and pharmaceutical purposes. However, many in society are concerned with how this technology might be extended to humans and what effects it might have on our environment. Maverick scientists aside, clones for the sake of self-aggrandisement are not on the agenda; nevertheless, the technology could be used to produce compatible organs for transplant and help infertile couples to have babies *without creating clones of either parent*.

The enactment of appropriate legislation regulating the bewildering array of possibilities offered by contemporary genetic manipulation requires enlightened debate. It is therefore a good time to reflect on how these extraordinary outcomes have been achieved; what limitations, if any, are involved in their application and where is it all leading us?

Acknowledgements

There are two people without whom this book would never have been written: Professor Vivian Moses who first suggested the project and whose patience and guidance ensured its fruition; and Dinky Manning who converted the early scribbled edition into beautifully word processed text, and also gave me encouragement with the project and advice on the content.

I also owe an enormous debt of gratitude to Finbarr Curran for his constant encouragement, coupled with fearlessly honest criticism; the book owes much to his advice. Many thanks are also due to numerous colleagues, relatives and friends: in particular Nat Khalawan for his continual support, advice and understanding; Professor Alan Hildrew for facilitating the project; John Curran, Denis Keating, Gerard Boyd, Mahua Chatterjee and Danielle Bugeja for advice and/or reading early drafts; and Rose and Jose Bugeja for providing invaluable support. My thanks also to the many individuals who contributed the illustrations that enrich the book and to Tracy Breakell and Jemma Nissel at Taylor and Francis, and Sarah Coulson at Wearset, for their consistent commitment to the project. Thanks also to Steve Turrington who did such a marvellous job as copy editor. Finally but most importantly I wish to thank my wife Virginia, whose love and patience brought me through many periods of self-doubt; she was ably assisted on many occasions by our little ones, Joseph and Francis, who constantly demonstrated that the frogs in the back garden were far more interesting than 'Dada's boring book with no pictures'. Now that the photographs have been added, I hope to be able to give the frogs at least a run for their money!

Uniquely similar!

You, dear reader, are unique. Nobody who has ever lived, or will ever live, is quite like you. Nobody will ever look exactly like you, think exactly like you, see the world with your eyes, hear the world with your ears, write with hand-writing that is quite like yours or indeed leave the same fingerprints as you do. And yet you, I, and everyone who has ever lived, are indistinguishable from one another in so many ways:

- we each began our lives as a single fertilised egg that became implanted in our mother's womb;
- a foetus developed and a new baby was born;
- helpless at first, but rapidly learning how to cope with the outside world, the baby became a young child, reached puberty and developed into a sexually mature individual who was then ready to start the life cycle of the next generation.

All our similarities notwithstanding, we are instantly distinguishable from one another and even from our closest relatives. I have a bump on my nose that I inherited from my father and short-sightedness that came from my maternal grandfather. My older brother looks like me (although he is by no means as handsome of course!). As his embarrassed girlfriends will tell you, my voice sounds identical to that of my younger brother over the telephone, yet be bears little physical resemblance to either of his two better-looking brothers (Figure 1.1). He has nevertheless an identical build to that of a paternal first cousin. His face may well resemble that of our mother but it is difficult to judge as she does not wear a beard. Suffice to say that clones we three are not!

But anyone who is reading this book might well be a clone; for every 350 readers, one will be. You will be physically almost indistinguishable from your brother (if you are a male) or sister (if you are a female). You will have similar likes and dislikes, taste in clothes and toiletries, be good at the same type of subject in school and probably even play the same sort of sport. Yet you will lead totally separate lives. Yes, human clones have existed ever since nature designed a system that sometimes produces identical twins.

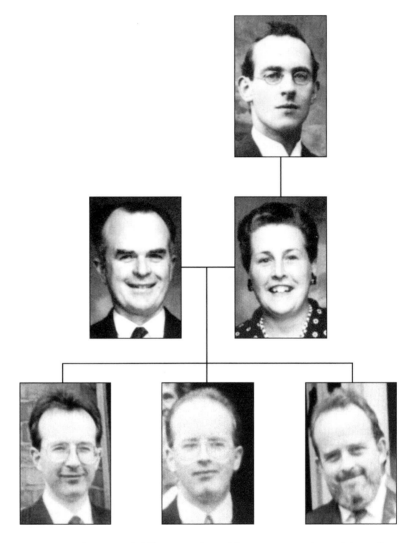

Figure 1.1 Similarities and differences inherited by various members of the author's family tree.

All humans begin life when a sperm cell from their father fuses with an egg cell from their mother. Non-identical twins arise when two different sperms fuse with two separate eggs at the same time, resulting in two off-spring (Figure 1.2a). These are as dissimilar as any singly conceived individuals and, indeed, are frequently of different sex. Human clones (identical twins) arise when a single fertilised egg divides to produce two cells, each one of which develops to produce a new individual (Figure 1.2b). United at conception, divided shortly afterwards, born as two independent individuals, these offspring are clones of one another. Animal clones of all shapes and sizes have existed for millennia, so why was there such a fuss when a cloned lamb called Dolly was born in 1997?

A clone with a difference

The birth of Dolly (Figure 1.3) heralded a revolution in reproductive techno-logy because she did not have a genetic mother or father but was instead 'propagated' by taking a piece of one sheep and treating it in such a way that it grew into a new identical animal. Strictly speaking, Dolly fails to fulfil any of the criteria for normal clones. She had no father, she was not con-ceived and (if you excluded the massed ranks of media vultures) Dolly was born alone. Impossible though it may seem, the animal with whom Dolly is identical is not her sister but her mother. Here we have a clone with a dif-ference – not produced as the result of a fertilised egg splitting to generate two separate individuals but by scientists taking an individual breast cell from a mature sheep and using biological tricks to grow a completely new identical animal from it. Rarely has a biological experiment been met with such hysteria from the press and genuine interest from the public at large who asked seriously:

- Why bother doing it?
- How does it work?
- Are human clones next?

A brief history of sex

For thousands of years, the collective wisdom on reproduction held that the female womb was a specialised environment to nourish and support the growth of a miniature baby implanted in it by the male sperm. Hence the biblical references to male 'seed' and 'barren womb' – analogies to plant horticulture which were well known at the time. This theory had many short-comings, not least of which were:

- Why were the offspring not all identical males?
- How did females arise?

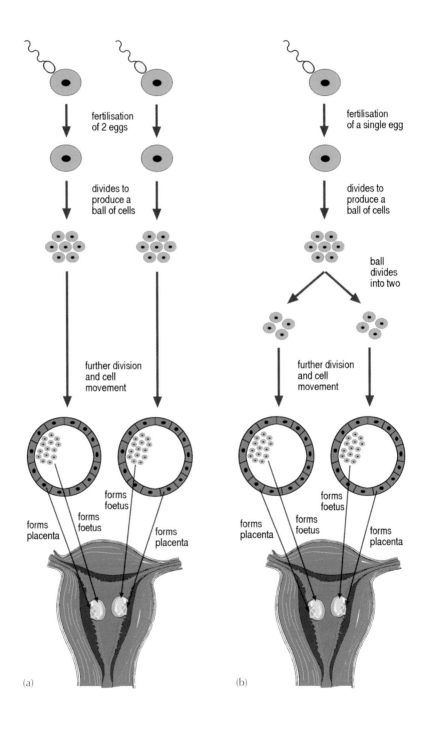

fertilisation
of a single egg

divides to
produce a
ball of cells

divides to
produce a
ball of cells

ball
divides
into two

further division
and cell
movement

further division
and cell
movement

forms
foetus

forms
foetus

forms
placenta

forms
foetus

forms
placenta

forms
placenta

forms
foetus

forms
placenta

(a)

(b)

Figure 1.3 Dolly was the first animal born without sexual reproduction. Reproduced
courtesy of the Roslin Institute.

- And how come that the offspring (male and female alike) often resembled
  their mother?

The predominately male thinkers of early history had little difficulty with
these shortcomings. Not surprisingly, later thinkers abandoned the theory of
a pre-packaged human planted in 'fertile soil' for one in which males and
females managed to co-operate and share in transmitting the *information*
needed to construct a new human being.

It was an obvious step but by no means an easy solution. Let us
suppose that the egg contains a copy of the information needed to construct
an animal identical with its mother while the male sperm possesses the

---

Figure 1.2 (a) Non-identical twins are the product of separate fertilisation events.
(b) If the ball of dividing cells in the early embryo splits in half, it can
produce two identical babies.

information for making an individual exactly like its father. Combining the two would produce an offspring carrying *all of the information from both parents*; it would have to be of *both* sexes and identical in all respects to all its siblings. Of course, neither of these things is actually true.

Even more impossibly, as each generation passed, progeny would accumulate twice the information content of their parents. Quite apart from the increasing burden of coping with so much data that the successive generations of progeny would contain, the fact that the original parents were viable with only the original amount of information testifies to the uselessness of multiple copies! No; somehow, not all of the information from *both* parents gets passed to their offspring.

As two individuals are involved in each act of conception, it makes sense to suggest that each contributes half of the information needed to assemble a new individual. But this, of course, cannot be the case because if it occurred at random, some progeny would have two noses and the other none, three ears versus one eye, etc. while if it took place in an organised fashion (e.g. nose from mother, eyes from father, and so on), all of the progeny would be identical with one another – each would have mother's nose, father's eyes and no other combination of characteristics would be possible.

No, nature has designed a much more subtle, foolproof system to allow each individual to exhibit traits that are different from both of its parents and all of its siblings. The same arrangement also prevents even a slight increase in the total amount of genetic information passed on from one generation to the next.

The solution is extremely elegant
Nature ensures that the information for each and every trait is inherited from *both* parents, thus ensuring that *two* copies of the recipe for *each* trait exist in their offspring. When the offspring themselves reproduce, nature provides a mechanism which again ensures that only one copy for each trait is passed on to the next generation (Figure 1.4). Even neater is the fact that the complete recipe for the animal passed on to the next generation is a random mixture of the two recipes in each of the parents. Such a random mixture from both parents generates unique individual progeny.

Figure 1.4 Nature provides two complete sets of assembly instructions only one of which gets passed on to the next generation. Like their parents each offspring possesses two complete sets of instructions but each one has a unique combination of traits.

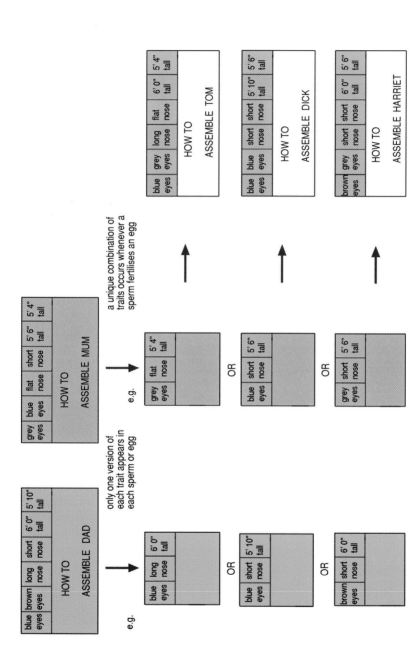

It is all done with sex – such a neat little trick that it is found throughout nature and for a very good reason. The genetic diversity that sex generates confers flexibility on the population as a whole by allowing it to adapt to changes in the environment; continual novelty among offspring makes it more likely that, even if the environment does change, some individuals will be able to cope and produce their own young. Animal breeders are familiar with this phenomenon. Few of the pioneers who bred wolves for use as domestic animals could have foreseen the resultant diversity of dog breeds available today. The genetic variation and flexibility found in the original wolf population has, through inbreeding, been channelled to generate different types of dog. Each type on its own constitutes a less varied, less flexible population of individuals – it is difficult to imagine breeding miniature poodles until they produced an Irish Wolfhound! The offspring from a cross between those types of dog would carry much more genetic variation than either of the parents and would almost certainly possess attributes shown by neither its mother or its father.

The problem with sex

Sex is designed to generate diversity, at which it is extremely good. Being unique might be very desirable for mankind but, for humans whose life's work is to breed prize herds of cattle, sheep and horses, the lottery of inherited traits that occurs every time an egg and sperm fuse is the last thing they desire. Inbred animals might all look very similar but there is still genetic variation within each individual breed – sufficient for an animal breeder to spend a lifetime breeding them for a uniquely desirable combination of traits. The offspring of a prize animal is rarely if ever as distinguished as its parents because it inherits only half its genetic information from its prize-winning mother or father. The other half comes from the second parent, and the same genetic lottery that gave rise to the unique combination of genes in the parent generates new diversity in the progeny. Even breeding from two prize-winning parents may not do the trick because the way the characteristics from each parent show themselves in the presence of the other is often highly unpredictable.

Quite simply, sexual reproduction will fail to produce uniform offspring. If, instead of relying on it, a magic wand could be waved over a prize dog to cause it to divide into ten new identical dogs, the problems caused by the lottery due to sexual reproduction would be avoided. The birth of Dolly the sheep shows that such a magic wand is now at hand.

No sex please – we're clones!

The word 'clone' is often associated with images from memorable science fiction movies where deranged scientists in darkened laboratories carry out sinister goings on. There is no doubt that 'real scientists' produce and work with a range of different types of clones, but nothing like the variety that nature herself generates.

The word 'clone' refers to one or more individuals that share identical genetic material. Clones abound in nature and, in fact, have been produced by a large section of the 'lay' population who have never darkened the door of a laboratory. Anyone who has grown an 'adult' plant from a spider-plant plantlet is a cloner! Propagating any plant from a cutting produces identical individuals – hence clones. But plants do not need man to help – strawberries (and many other plants) produce runners giving rise to new plants which are clones of the original. 'I must have a cutting!' is synonymous with 'I would like to have a plant identical in every way to the one you have!' There isn't a gardener in the country who has not cloned some plant or other.

But clones are not restricted to plants in nature. The many yeast strains used to produce different types of beer are examples of yeast cell clones. The *E. coli* outbreak in Scotland in 1997 was caused by a particular strain of bacteria, an example of a bacterial clone. The different strains of influenza virus are examples of viral clones, each one requiring a separate vaccine. A single cancer cell can produce a tumour which is a ball of identical cells – clones of the original. Clones are simply a group of individuals possessing identical genetic information. The majority of naturally occurring clones arise without any need of sex, but animal clones occur in nature only as identical twins (or multiple births) who began their life journey as one single fertilised egg which divided to give separate and independent individuals at birth. Animal clones are special because they are the direct result of sexual reproduction; all the clones of plants or microbes come from simple cell division and continued growth as new individuals.

Sex produces diversity; clones produce uniformity. Nature values both but, on balance, prefers sex. Mankind may enjoy the latter but on balance prefers clones when it comes to earning his living. Agriculture is based on uniformity. Where uniform clones are available they are used; until the 1970s the entire South American coffee industry was based on one cutting from a single coffee plant which had been maintained at Kew Gardens in London during the 1800s. Where clones are unavailable, plant and animal breeders use intensive breeding programmes in order to generate the desired uniformity. Prize herds of animals are created by artificial insemination programmes. In the Philippines 70% of all rice planted, and 90% of that in Columbia, is a specific high-yielding dwarf variety. There is a similar story for wheat in Mexico and India.

Propagating animals without sex

Nearly every cell in a multi-cellular organism contains the entire genetic blueprint necessary to produce that organism. In theory, therefore, every one of those cells should be capable of growing into a new individual. In the early stages of embryo development in an animal this is indeed correct – hence the birth of identical twins, one from each of the two cells formed by the first cell division after fertilisation. However, as the embryo grows and develops different cell types, this facility disappears; but with many plants it does remain possible to separate out single cells, grow them in special nutrient conditions and get them to divide into normal plants – clones of the original.

So far this has not been achieved with animals because there seems to be something about an animal egg cell which is uniquely necessary for an embryo to develop. Nevertheless, as far back as the 1960s, scientists were able to show that mature adult cells contained the genetic blueprint for the entire animal. Removing the nucleus from a fertilised frog egg cell and replacing it with one from a frog skin cell, they succeed in getting tadpole development. These tadpoles were, of course, clones because their genetic material came from a single skin cell and not as a result of sexual reproduction. Attempts to repeat this type of experiment with warm-blooded animals (mammals) failed and, in fact, such unsuccessful bids to clone mice prompted one scientist to remark as recently as 1986 that 'the cloning of mammals, by simple nuclear transfer, is biologically impossible'. How wrong he was.

Enter Dolly – born only 12 years later, she reminded us once again of the power that can be released when scientific ingenuity is applied to nature's incredible flexibility. The diagrams in Figure 1.5 illustrate this process. In sexual reproduction, the fertilisation of individual eggs by different sperm generates a range of different individuals. In animal cloning these fertilised eggs have their nuclei removed and each one is replaced by a nucleus taken from donor cells, all of which originated in one animal. If these develop into progeny after being transferred into the womb of a surrogate mother, each one will have a constitution identical genetically with that of the donor animal and they will also be identical to one another. In Dolly's case, the precise details were that cells were taken from the udder of a 6-year-old Finn Dorset sheep and grown for a few days in special nutrient

Figure 1.5 The 'instruction manual' for the assembly of Dolly was a tried and tested one from another animal; not a random composite provided by the fertilisation of an egg by a sperm.

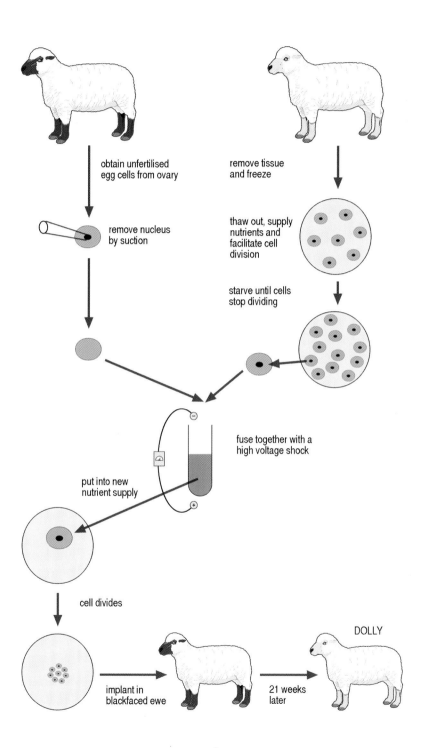

medium, during which time the nutrient was slowly removed so the cells became starved and entered into an inactive state. Then, mature egg cells were taken from Scottish Blackface ewes and their nuclei removed. The nuclei were excised from the inactive udder cells and fused with the enucleated egg cells using an electrical pulse temporarily to punch holes in the egg cell membrane big enough to allow the new nucleus in. The electric pulse also 'tricks the egg into believing' that it has been fertilised by a sperm and so prompts it to begin dividing and developing into an embryo (Figure 1.5). The cells were monitored for a few days to ensure that cell division had started before they were transferred to the wombs of surrogate mothers who were to carry them to term. Of 277 that were fused, 29 started to form an embryo but of these only one resulted in the birth of a lamb.

Dolly was unique in the history of animal husbandry because she had been 'propagated' from another individual without the use of sexual reproduction, just like a plant cutting. Furthermore, she carried exactly the same genetic information as her predecessor despite the fact that she did not share its parents. Scientists had found a way to avoid the lottery of sexual reproduction by developing procedures allowing them to obtain *both* recipe books from an adult animal and implant them in a prepared womb. This means that, in theory at any rate, as many identical progeny as one required could be generated in this way from a single animal. The womb in which the foetus develops does not have to be related to the future offspring. Seeds and fertile wombs spring back to mind.

Animal breeding aside, the real significance of Dolly was that she was an unequivocal answer to a basic biological question long asked by scientists. We knew that all of the information to make an animal is contained in the fertilised egg at its conception. Dolly's birth revealed that all of that information is still safely contained in the nucleus of adult cells. Some 25 years earlier scientists had learned how to manipulate small fractions of this total information; now they could manipulate all of it. More than two decades ago they had unlocked Pandora's box; now it had been thrown wide open.

One thousand separate nuclei would fit on the full stop at the end of this sentence. How is such a huge amount of information carried in so minute a volume? How can all of that information still be present in adult cells and how does it manifest itself in complex creatures like Dolly or you or me? To address this question we must embark on an amazing journey, one made even more fascinating by approaching it from a human perspective.

2

Incredible journeys

'One small step for man. A giant leap for mankind.' Those famous words spoken by Neil Armstrong as he became the first man to walk on the moon induced unforgettable scenes of excitement at mission control in Houston. The teams of hundreds of specialist scientists and engineers had achieved this incredible feat by harnessing the amazing power generated when specialisation, communication and co-operation are used to solve a complex and difficult task. Your body consists of 50 thousand billion cells, yet not so long ago you were just a single fertilised egg. Much more impressive than a human going to the moon is a human going from a single cell to a sentient being capable of rational thought.

Wondrously made

Look at your skin – to the naked eye a smooth, soft, extremely tough covering; under a microscope millions of individual cells (Figure 2.1). Cut the skin and a fluid oozes out – red to the naked eye; containing millions of curiously shaped red cells under the microscope (Figure 2.2).

Take a deep breath and hold it for as long as you can. How long did you manage? Thirty seconds to 1 minute, average; 1–2 minutes, good; 2–3 minutes, you must play a trumpet. More than 3 minutes and I hope an appreciative relative will get this book in your will! All animal cells need oxygen; some cells need it more than others do – muscle cells can do without it for limited periods but nerve cells need it continually. As a professional strangler will tell you, stop the supply of blood to the brain for 1–2 minutes and your next victim is in hand. Blood is necessary if oxygen is to get to the cells but, unlike a sugar such as glucose, oxygen does not dissolve very well in water, so specialised red blood cells do the job. They pick up oxygen in the lungs and get pumped by the heart via the arteries all around the body. Having delivered their critical cargo of oxygen to body cells, they return via the veins and heart to the lungs for more oxygen. The act of carrying oxygen actually alters the colour of the cells: arterial blood is scarlet and venous blood is dark red.

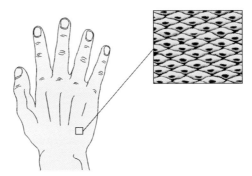

Figure 2.1 Skin may appear to be a single continuous layer but it consists of millions of cells connected to one another.

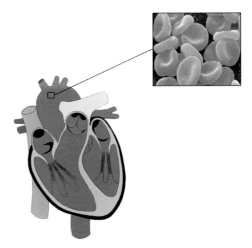

Figure 2.2 The heart pumps blood to every cell in the body. Under the microscope, blood consists of thousands of discrete cells of different types. The red ones carry the oxygen.

Cells also need energy. Approximately half the blood volume is slightly salted water, carrying all sorts of goodies around the body. This liquid is the source of nutrient for all of the 50 trillion cells in the body. The blood obtains its nutrient load from the intestines that we reload every time we eat. But we eat all sorts of complex food and our cells really need glucose and a number of other relatively simple chemicals. It is for this reason that the entire intestinal tract is dedicated to converting pizzas, hamburgers, fries and caviar into that chemical food that the body cells can assimilate.

In the course of assimilating nutrients and assembling cellular components, all cells produce toxic side products. Human bodies accumulate huge amounts of all types of toxins both ingested with and derived from processing complex foods (not to mention alcohol!), and as natural by-products of our cells' normal activities. Our blood removes the toxins from cells throughout the body and delivers them to the liver. There a whole battery of processes alters the toxins and makes sure that they are detoxified as much as possible before being returned to the blood for carriage to the kidneys and excretion in the urine.

All the while, our bodies have to interact with the external environment. The control and command centre in our brain gathers information with our senses, as other sensors monitor innumerable internal conditions, and helps the body to respond as a unified whole. Bend your finger – how did that happen? Your eyes are picking up light reflected from this page you are reading and signals on the retina at the back of the eye are transmitted by nerve cells to the mass of nerve cells, your brain. There the information is processed, used to make a decision which is passed on to a centre controlling finger movement and communicated via more nerve cells to muscle cells in the hand, telling them to twitch in one direction – and to do so all together. The muscles are joined to bone by ligaments and so your finger moves.

Even an emotion like fear sends chemicals surging through the brain and nervous systems, flooding into the bloodstream, communicating with muscles to move limbs, stimulating the heart to speed up and the blood pressure to rise so we are ready to fight (or run) to protect our whole organism. Less dramatically, and over a longer time-scale, babies grow into adults and thus require the co-ordinated development of bones, organs and supply systems. When athletes train at high altitude, the body senses a drop in oxygen concentrations in the blood. It sends a 'messenger' coursing through the blood to tell the bone marrow to increase the number of new blood cells and allow what little oxygen there is to be trapped more efficiently in the lungs.

The 50 trillion cells that arose from a fertilised egg have undergone such an astonishing strategy of specialisation, communication and co-operation to produce a being with a degree of complexity that makes rockets and space shuttles look like childrens' toys.

A cell is a cell is a cell

The vast majority of cells are so small that they can be seen only by using a microscope. Their physical form is like that of a tiny soap bubble – a thin flexible membrane separating the internal environment from the external one. Just as a soap bubble's shape can be altered, so too do cells assume many different contours. Unlike soap bubbles, which are both filled with air and floating in air, the internal contents of cells are dramatically different from their external surroundings. The internal contents are a complex mixture of millions of tiny components interacting with one another to maintain the integrity of the cell. So different is their inside from the world outside that they are completely covered by a very protective membrane to keep the outside out and the inside in.

Cells have a number of basic functions (Figure 2.3):

1. They assimilate nutrients from the environment.
2. They use them to assemble the various components that are in the cell.
3. They eliminate waste products.
4. They grow bigger until they reach a critical size and *then the cell divides into two.*
5. Each one of these daughter cells can take up nutrients and begin this cycle all over again.

Some organisms (like bacteria and yeast) spend their lives as single cells and do little more than go round and round this circus of events (Figure 2.3). The cells in a multi-cellular organism such as a human also carry out these basic functions but, in addition, they specialise: blood cells, skin cells, muscle cells, nerve cells and all the other varieties have recognisably different physical forms and thus manifest different 'assembly instructions'. Remember that each one of us humans arose from a single microscopically small fertilised egg and that all the information necessary for the development of our bodies from then until now resided in its nucleus. Controlled by this information, the cells which developed as that fertilised egg divided assimilated nutrients, assembled components, grew bigger and continued to

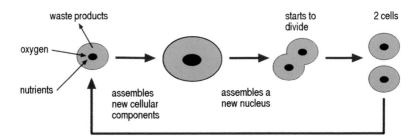

Figure 2.3 Cells grow bigger by fashioning nutrients from the environment into a huge array of cellular components. They then divide to produce two new cells.

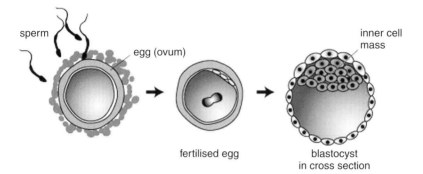

sperm

egg (ovum)

inner cell mass

fertilised egg

blastocyst in cross section

Figure 2.4 All of the cells in the early embryo of a multi-celled organism like man look the same and they execute the functions of simple cells. Reproduced courtesy of National Institute of Health USA.

divide (Figure 2.4). Even as the ball of cells increased in size, some of them started to specialise and assume a variety of different shapes and sizes, ultimately to form the various tissues and then the organs which are still maintained within our bodies today.

Misadventures
In 1975 the world held its breath for 3 days as the crew of Apollo 13 struggled to return their stricken vessel to earth. In 1993 the world's media carried dramatic pictures of Challenger 2 as it exploded shortly after take off with the loss of all on board. These provide startling evidence, if any is needed, that incredible journeys started in hope can so easily end in disaster.

In many ways, complexity inspires awe because it continually resists the natural tendency of things to descend into chaos. But chaos is never too far away: the Challenger space shuttle exploded simply because, despite the millions of components that were correctly assembled, a defective O-ring went undetected. Some misadventures are more serious than others. The crew of Apollo 13 escaped with their lives; those in Challenger were not so lucky. Living organisms are prime examples of complex systems and they, too, suffer if one or more of their components fail to work correctly. Some failures they can live with; others are lethal.

Cellular misadventures
The normal shape of a red blood cell, one that picks up oxygen in the lungs and transports it to the body cells, is that of a biconcave disc (Figure 2.5 (a)). This shape is necessary because it allows many to pack closely together

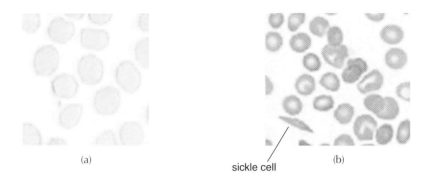

(a)

sickle cell

(b)

Figure 2.5 When the normal rounded shape of red blood cells (a) becomes distorted
(b) the affected individual develops severe anaemia. Reproduced courtesy
of QMUL Medical School.

without clogging the very fine blood vessels (called capillaries) that deliver
the blood to the organs all over the body. There is, however, a condition
called *sickle cell anaemia* in which the red cells have a fundamentally different shape (Figure 2.5 (b)); this minor alteration in just one type of body cell
has a dramatic knock-on effect on the entire organism. In this disease, for
that is what it is, the cells have a crescent or sickle shape which results in
their becoming jammed in fine blood vessels and in clumping together. This
causes blockage of the blood supply to all the major organs with consequent
damage to lungs, kidneys, heart, intestines and brain. The body recognises
that they are the wrong shape and traps them in the spleen to destroy them.
Unfortunately this further aggravates matters because the person then develops anaemia (or lack of red blood cells), putting even more pressure on the
already compromised heart. If left untreated, a person suffering from this condition will die at a very young age. Even with regular blood transfusions, the
patient's quality of life is severely constrained.

One of the major functions of the cells in the skin is to secrete water in
order to cool the body. If a drop of perspiration ever dropped onto your
tongue you probably noticed that it tasted salty. That is because some salt
escapes with the water as it is pumped out of the body. However, salt is vital
to the body's well-being and the skin cells, and their close relatives in the
pancreas, lungs and reproductive tracts, put lots of energy into reclaiming as
much salt as possible before the water is finally released. Just how important
the reabsorption of salt is to the body becomes apparent when these cells fail
to do it (Figure 2.6). When the cells secrete water containing high levels of
salt, this not only makes perspiration very salty but also has a devastating

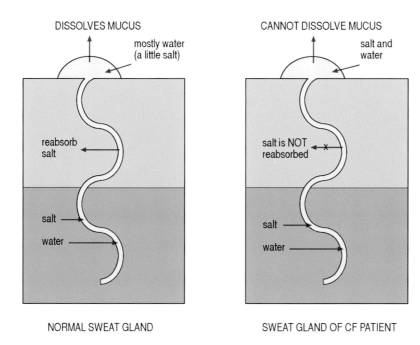

Figure 2.6 The inability of a number of cell types to reabsorb salt causes mucus to accumulate and cause extensive problems for the affected individuals.

effect on the internal organs. Too much salt in the watery secretions prevents them from properly dissolving the mucus associated with the cells in the internal organs. The mucus blocks the ducts which drain digestive juices into the intestines from the pancreas. These juices accumulate and start to digest the tissue that produced them, so damaging the pancreas. Sterility results from blocked or otherwise damaged tubes in the reproductive tract. But the most devastating effect is that wrought on the lungs, where the viscous mucus provides an ideal breeding ground for all sorts of bacteria, leading to chronic chest infections.

Until recently, babies born with this defect did not survive to their first birthday. Today, thankfully, such individuals have a much greater life expectancy due to a combination of therapies such as antibiotics to treat infections, chest physiotherapy to dislodge the mucus and ingested digestive enzymes to replace those missing due to pancreatic damage. In its severest form, however, this aggressive disease, still tragically curtails far too many lives. It is of course the disease we call *cystic fibrosis.*

Sickle cell anaemia and cystic fibrosis are two simple examples of what can happen with the appearance of an apparently slight change in just *one* cell type in an entire organism. Given that each cell consists of thousands of different components the question now becomes: which one is to blame?

A musical interlude

Whatever your taste in music, it is probably fair to say that different pieces have very different effects on you. Some are relaxing, others excite, elevate, depress or agitate. Some works are short, others long; some are simple, others complicated; yet all are composed by simply arranging the members of a small specific set of individual notes into different patterns. It is the relative position and length of each note with respect to all the others that determines the specific structure of a piece of music. It is this structure in turn that elicits an effect on the listener.

The complex harmony present in multi-cellular organisms arises from the interaction of innumerable members of a special type of biological art form, one which has a huge diversity of structures. Some are simple, some complex, some small and some large but all are composed of the same *twenty* sub-units or building blocks called *amino-acids* (Figure 2.7). Designed to interlock with one another, these twenty molecules (see Box 2.1), each with its own shape and characteristics, are combined into the countless different, yet quite specific biological art forms we call *proteins* (Figure 2.8).

BOX 2.1 All you ever wanted to know about molecules but were afraid to ask

All matter from the stars in the sky to the hair on our heads consists of trillions of incredibly tiny particles called *atoms*. Atoms come in a variety of shapes and sizes (more than 100 in all) and they can join with one another only in very specific ways to produce combinations of atoms held together in what are called *molecules*. The shape of a molecule is determined by the atoms which comprise it. Each molecule is a specific size and shape and therefore has its own specific charcteristics. The water we swim in, wash with and drink consists of molecules of H_2O: two atoms of hydrogen joined to one of oxygen. If the two hydrogen atoms are replaced with one oxygen atom the molecule becomes one of oxygen (O_2) – the gas we breathe in. If one atom of carbon is inserted between the two oxygen atoms, a carbon dioxide molecule (CO_2) is formed – the gas we breathe out.

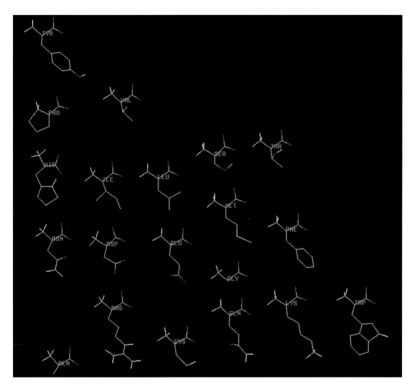

Figure 2.7 Each amino-acid has its own specific characteristic shape and function. 'IMB Jena Image Library of Biological Macromolecules (www.imb-jena. de/image.html)'

Add three hydrogens to one atom of nitrogen and a molecule of ammonia (NH_3) is formed – this is the pungent stuff in smelling salts, the active ingredient in many household cleaners and it is also found in the fertiliser we put on our gardens to supply plants with nitrogen. Replace the two hydrogens in water with two nitrogens and a molecule called nitrous oxide (N_2O) is formed. This is commonly known as 'laughing gas', found in the dentist's surgery that acts as an anaesthetic. Suffice to say that the atoms that go to make up a molecule determine its characteristics.

From the smallest flu virus to the giant redwood trees to man – 99% of living matter consists of only C, N, O and H! The incredible diversity of living organisms arises because carbon, which can join with four other atoms, 'likes' to join with other carbon atoms and so create long chains. Each additional carbon creates a different molecule and each new carbon in a chain can join with hydrogen, nitrogen, oxygen or more carbon to produce an endless variety of possible shapes and sizes for nature to play with. When two carbons, six hydrogens and one oxygen are joined together the molecule is alchohol. Drop off one hydrogen and the molecule becomes vinegar – thus explaining how easily wine can become vinegar!

Nature also likes to make groups of related yet subtly different molecules which are designed to lock and unlock to one another easily. These are used as building modules for the rapid assembly and disassembly of giant molecules consisting of hundreds, thousands or even millions of atoms all linked together. The amino acids are one such group.

Proteins execute every one of the tens of thousands of different cellular functions associated with the trillions of cells found within a human body. Some, such as those found in muscle and hair, play a mechanical or structural role. Others called *enzymes*, of which there are 3,000 or more in humans, specialise in assembling or digesting other molecules. Some proteins sit like satellite dishes in cell membranes where they receive signals in the shape of yet other proteins called *hormones* which dictate how the cells are to behave (see Box 2.2). In short, the functions that a cell can execute are a direct con-

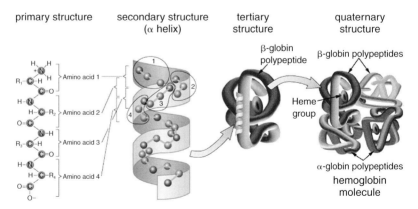

Figure 2.8 Amino-acids interact with one another conferring a specific shape on the resulting protein molecule. Reproduced courtesy of National Institute of Health USA.

Box 2.2 Hormones make our cells go around!

Hormones are uniquely shaped homing devices used by cells to communicate with one another. One cell produces the hormone and releases it into the blood where it circulates. The 'target cells' have specific receptor proteins in their external membrane which act a bit like car ignitions – when the correct hormone inserts itself into the receptor, the cellular engine gets switched on. Cells lacking the correct receptor remain unmoved by the presence of the hormone. Hormones need to have a very specific size and shape and therefore many of them are proteins (Figure 2.9)

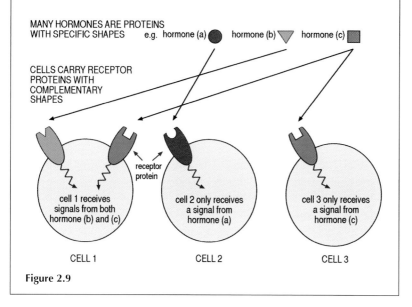

Figure 2.9

sequence of the proteins that it contains. Some proteins are found in every cell; others only in specialised cells. For example, the many proteins needed to degrade glucose in order to release its energy are in all cell types yet the *haemoglobin* protein, which carries oxygen around the body, is present only in red blood cells.

The digestion and assimilation of food are further examples of specialised protein activity. A whole series of specialised cells secrete a selection of different enzyme proteins into the intestines to help digest our food: amylase digests starch, trypsin and chymotrypsin digest proteins and lipases digest lipids. As the blood becomes laden with 'goodies', messengers called *hormones* (see Box 2.2) ensure that the body assimilates them. *Insulin* is secreted into the blood following a meal and lowers blood glucose levels by binding to insulin *receptor* proteins in liver cells, causing them to store glucose for future use. When the blood glucose level is too low, yet another hormone protein, called *glucagon*, stimulates the liver cells to release the stored glucose back into the bloodstream. Hormones also govern body size. *Growth hormone*, a protein that sometimes features in the news as a banned substance abused by some athletes in search of extra muscle bulk, stimulates bone, liver, muscle and many other cells to divide.

Muscle cells contain specialised proteins (*actin* and *myosin*) which combine to produce incredibly strong fibres. Thus, thigh (and any other) muscle, whether from a frog or an elephant, provides a source of protein nourishment. Another fibrous protein (*keratin*) forms the hair on your head and the nails on your fingers and toes. Finally, every cellular component (including the proteins themselves) are themselves assembled by proteins, while the process of cell division is initiated, mediated and stopped by an army of them, each with its unique sequence of amino acids, each with its own critical role to play in this most critical of processes.

No class of biological chemicals is more important; proteins govern each and every one of the roles performed in each and every cell in our bodies. The incredible similarities seen in the faces of identical twins reveal that their biological sculpting tools (each one a protein) are identical in every respect in each of the twins. Equally, despite the same basic facial architecture, the very obvious differences seen in the faces of non-identical twins reveal the biological tools used to sculpt these individuals as similar but by no means identical in every respect.

Discordant notes

The precise structure of a piece of music is determined by the order of the individual notes of which it is comprised. So too, a precise structure and hence function is conferred on each of the estimated 120,000 different types of proteins in a human body by the unique order of its constituent amino-acids. Moreover, just as an incorrect note may or may not jeopardise the integrity of a piece of music, an incorrect amino-acid sequence may or may not jeopardise the structure and hence the function of a protein (Figure 2.10). Some changes go unnoticed; some, while noticeable, can be endured; while others can result in a loss of structure and with it a cascade of events culminating in a disastrous case of biological dissonance.

Cells become specialised because they contain proteins whose structure confers a specialised function on them. So, too, cells become *compromised* when one or more of the proteins that they contain are assembled incorrectly, causing a change in the structure and hence function of the proteins. Indeed, the defective cells mentioned in the previous section all arose as a result of just such minuscule changes in amino-acid sequence.

Dysfunctional proteins

Each red blood cell contains tens of thousands of molecules of the oxygen-carrying protein haemoglobin. When the order of these amino-acid sub-units in haemoglobin from normal and sickle-shaped red blood cells was examined, a slight difference was found in their amino-acid sequences. The protein from the normally shaped cell had an amino-acid at position 6 in the chain able to interact with water; the amino-acid at the corresponding position in haemoglobin from sickle cells was a sub-unit that avoided water as much as possible. Just as a drop of oil on the surface of water gathers itself into a tight ball so, too, the shape of the sickle-cell protein at this position has been changed. This minor structural alteration causes the defective protein to join up rapidly with similarly defective haemoglobin proteins; instead of each protein remaining apart, they coalesce to form long fatty chains of haemoglobin (Figure 2.10 (A)). Such structures distort the red blood cell into a sickle shape with severe consequences for the sufferer unless he or she receives urgent medical attention. Despite the fact that all of the other 286 amino-acids in the sufferer's haemoglobin molecules are correct, and despite the fact that *all* of the tens of thousands of other proteins present in that individual are working properly, the alteration in protein structure and function is devastating.

A protein defect also lies at the heart of cystic fibrosis. We have seen that this disease arises when the channels conducting water beyond the layer of epithelial cells secrete fluid carrying too much salt into the internal environment. The external membrane of all cells, including epithelial ones, is an extremely complex structure consisting of fats and scores of proteins carrying out numerous functions. Some of the structures, made up of amino-acids, are minuscule pores in the membranes through which chemicals such as salt can pass. Normal cells have a channel-forming protein (1,480 amino-acids long) which is totally absent from the membranes of similar cells from cystic fibrosis sufferers. The damaged cells of cystic fibrosis patients nevertheless do produce an aberrant protein which simply lacks the amino-acid found at position 508 in the normal version. This minor alteration prevents the protein from being inserted into the membrane, thus causing the essential channel to be absent (Figure 2.10 (B)). As with sickle cell anaemia, despite the fact that all of the other proteins in the body are functioning correctly, a change of one amino-acid out of 1,480 in a large protein is sufficient to interfere with its salt-secreting function and wreak havoc on the entire organism.

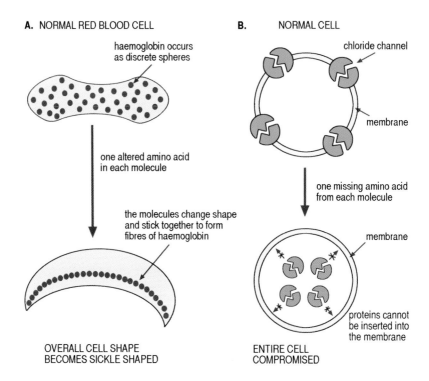

A. NORMAL RED BLOOD CELL

haemoglobin occurs
as discrete spheres

one altered amino acid
in each molecule

the molecules change shape
and stick together to form
fibres of haemoglobin

OVERALL CELL SHAPE
BECOMES SICKLE SHAPED

B. NORMAL CELL

chloride channel

membrane

one missing amino acid
from each molecule

membrane

proteins cannot
be inserted into
the membrane

ENTIRE CELL
COMPROMISED

Figure 2.10 (a, b) An altered amino acid can affect the overall structure of a protein.

The Dolly experiment shows that the information for the assembly of an entire animal resides in the nucleus of each cell. For that reason it must also be the repository of any defective information that an organism might manifest. Therefore it is to this structure that we must take our search if we are to find the reason why organisms can go so badly wrong.

3

What is it about the nucleus?

If you use a reasonably strong microscope to look at a nucleus from a human cell which has been stained with special dyes, you will be able to see structures called *chromosomes*, of which there are 46 in human cells (Figure 3.1).

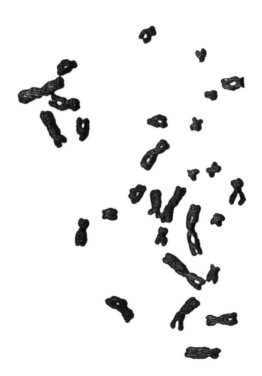

Figure 3.1 The 46 chromosomes found in human cells can be arranged into two sets of 23. One set we inherited from Mum; the other from Dad.

The chromosomes are rather like the animals in Noah's ark: all different shapes but arranged in pairs. One member of each of the 23 pairs we inherited from mum, the other from dad. We know that the assembly of a human (or any other large organism) requires tens of thousands of separate instructions: each chromosome must therefore carry the information for many thousands of different functions. Each nucleus (excluding those found in the sperm or egg cells) contains a complete set of chromosomes and plays a vital role as the repository of information required for the assembly and maintenance of each and every cell in the body. What is the source of all that information and how does it mediate the assembly of cells and entire individuals?

Chromosomes and gender

There is something special about one pair of chromosomes. In 22 out of the 23 pairs, the two chromosomes (one originally from mother, the other from father), while not identical, are certainly very similar. They are so similar in fact that just before they get passed on to the next generation, they actually interchange bits between themselves so that the individual chromosomes in the sperm or egg represent a mixture of both maternal and paternal lineages (Figure 3.2). But the twenty-third set, called 'X' and 'Y', are so different that almost no exchange can take place. Females have two Xs in the twenty-third pair (they can undergo exchange) and each of their eggs therefore has just one X. Males, however, carry one X plus one Y; as sperm have only one or the other, about half of a man's sperm are X and half Y. If an egg is fertilised with an X sperm the resulting individual is XX and female; if with a Y sperm, the combination is XY and hence a male. That is how boys and girls result from the chance mating of an egg with one or other type of sperm (Figure 3.3).

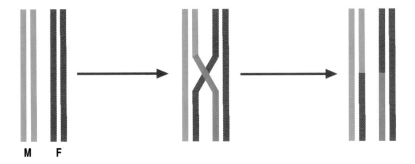

M   F

Figure 3.2 The chromosomes in cells destined to produce sperms or eggs can swap pieces with one another.

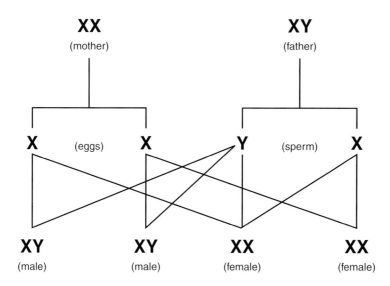

Figure 3.3 If a fertilised egg contains a Y chromosome it will develop into a male foetus.

The long and winding code

It is possible using modern high tech procedures to undertake a more detailed analysis of chromosomes. This shows that they contain metres (yes, metres!) of a substance called *DNA* (*deoxyribonucleic acid*), all wrapped extremely tightly around proteins acting like cotton reels (Figure 3.4). They stop the molecule getting tangled and also seem to protect the DNA from being damaged during the manoeuvres in which it must become involved during cell division. Such compression of long stretches of DNA into a single nucleus (a microdot, as it were) is possible only because the DNA molecule is infinitesimally thin. Incredibly thin, incredibly long and incredibly organised in the nucleus: no verbal or visual aid can ever adequately describe the DNA molecule.

It is just as difficult to explain two very remarkable facts:

1. No matter from which organism DNA comes (human, sheep, fly, yeast, bacteria, virus), it always has the same physical structure – that of a double helix (Figure 3.5) – although its detailed structure varies in a very subtle way which we will come to quite soon.
2. Breaking DNA molecules down into their component parts shows that the chemical building blocks are also always the same: a sugar (called

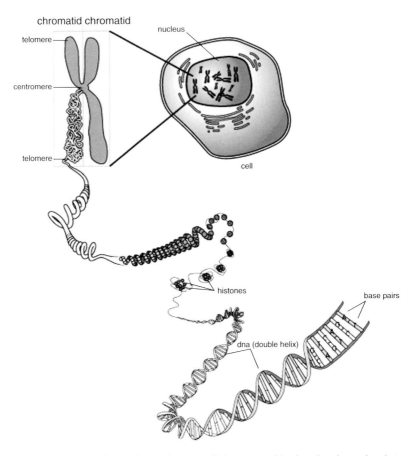

Figure 3.4 Metres of DNA fit into human cells by virtue of the fact that the molecule is infinitesimally thin and extensively packaged. Reproduced courtesy of National Institutue of Health USA.

deoxyribose), four basic chemical structures, called *bases*, whose full chemical names and abbreviations are *adenine* (A), *thymine* (T), *guanine* (G) and *cytosine* (C), and a *phosphate* backbone (Figure 3.6).

The basic simplicity of DNA molecules reflects the fact that they carry out relatively simple yet absolutely essential functions; the universality of its structure is a reflection of the fact that it carries out the same functions in every organism in which it is found:

(a)

(b)

cytosine   guanine

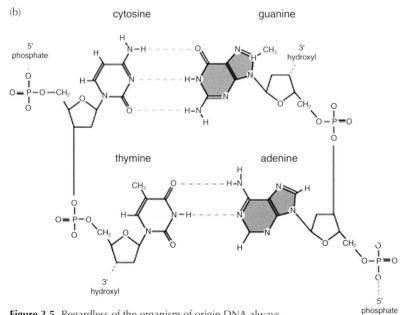

Figure 3.5 Regardless of the organism of origin DNA always
(a) Has the same physical appearance.
(b) Consists of six basic chemical components.
Reproduced courtesy of National Institute of Health USA.

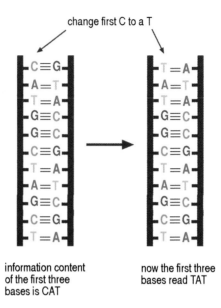

Figure 3.6 The order of the bases along one strand constitutes the information content of any given DNA molecule.

- to encode all the information necessary for the growth and development of the organism that contains it, and
- to facilitate the transmission of that information to future generations of cells and progeny.

To understand how it does so we must prise the double helix apart and peer inside. If you could grab hold of both strands of the helix and pull them apart they would separate just like two strands in a piece of string. Release your grip and, unlike the string that reluctantly attempts to go back to its original structure, the two strands of DNA snap back immediately into precisely the same position they had before you pulled them apart. They do that because a number of tiny 'magnets' in critical positions on the inside faces of the bases on either strand of the DNA helix attract one another in a very specific fashion. A attracts T through two 'magnets' set a precise distance apart, whereas G attracts C through three 'magnets' set at a different distance apart (Figure 3.5 (b)). This ensures that A only ever joins T whereas G only ever joins up with C.

If we prise the two strands apart a second time and hold them apart long enough to scan down the length of one strand we will see an endless string of bases all joined together in an apparently random combination e.g. … TATGGCTAG … . If we now look at the sequence of bases on the opposite strand, it reads … ATACCGATC … because $^T_A{}^A_T{}^T_A{}^G_C{}^G_C{}^C_G{}^T_A{}^A_T{}^G_C$ was how it was organised before we pulled it apart (Figure 3.6), all due to the very precise and specific A-T and G-C pairing; base pairing means that each strand is *complementary* to the other.

Now we can see how such an apparently simple structure can achieve the two functions of code and transmission: the *order of bases on one strand is itself the code* (Figure 3.6). As you might expect, a comparison of this order for different types of organism reveals huge differences, because the details of the code for humans is obviously not going to be the same as those for snails or bananas. The accurate and exact transmission of this information from one generation of cells to the next is achieved by a mechanism that unzips the DNA and assembles new strands against those of the separated helix. As A joins only to T, and G joins only to C, the sequence of the two *new* double helices is identical both to the original and to one another (Figure 3.7). This amazingly simple yet beautiful solution to information transfer is reasonably intuitive; how the order of four bases along the DNA can hold all of the information required for the assembly of an entire organism is less so.

We have already seen that proteins determine cell organisation and function while their own properties derive directly from the precise order of

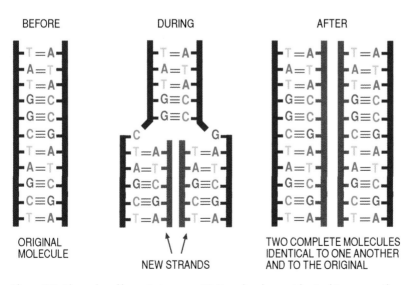

Figure 3.7 The order of bases in two new DNA molecules are identical to one another and to the original.

their constituent amino-acids. The simplest way for DNA to control an organism is by specifying the order of amino-acids in its proteins; that will determine the structure, and hence the function, of each protein molecule. But there are only four different DNA bases (A, T, G and C) as against twenty different amino-acids to be found in proteins; how can the order of four bases in the DNA determine the sequence of twenty amino-acids in the proteins? The answer is, by means of a code.

The code

The Morse code has only two symbols – a dot or a dash – yet it codes for all 26 letters as well as for numbers by using groups of dots and dashes read together to symbolise the different letters and numbers. The genetic code works on the same principle. The sequence of DNA bases is read in groups of *three* and each triplet is called a *codon*. The first position can be occupied by any one of the four bases A, T, G or C; there are similar choices for both the second and third positions giving a total of $4 \times 4 \times 4 = 64$ *different* combinations (a lot more than Morse could manage with his two symbols). All of these combinations, and the amino-acids they encode, are listed in Table 3.1. Note that the genetic code also includes punctuation; three codons out of the total of 64 specify 'stop'. As the other 61 combinations specify only twenty amino-acids, all the amino-acids in protein except two are coded for by more than one triplet, some by as many as six. The reasons for this variability are buried deep in biological history; some people think that the amino-acids with many codons were components of protein right at the beginning of biological development, when the DNA-protein system first evolved, while others came later and had to make do with fewer triplet codons.

Now we can appreciate just what a *gene* is: a segment of information in DNA that specifies the structure and function of a particular protein by using a triplet code to determine the order of its constituent amino-acids.

The missing link

Humans have approximately 75,000 different genes, encoding the structure and hence function of the tens of thousands of different proteins found in the body. Each gene consists of a unique sequence of bases somewhere along one of the enormously long DNA molecules that are condensed into the chromosomes found *inside* the nucleus. The encoded proteins, on the hand, are assembled *outside* the nucleus in the cytoplasm of the cells by large numbers of sub-microscopic units called *ribosomes*. A mechanism therefore exists to transfer the genetic information from the DNA in the nucleus to the ribosomes in the cytoplasm.

Table 3.1 This table contains every possible combination of A, T, G, C taking them 3 at a time. The encoded amino-acids are also presented. Adapted from *Genetics* Judith Bunting Publications, Boxtree 1994 by Glenn Smith

1st Position ↓	2nd Position				3rd Position ↓
	T	**C**	**A**	**G**	
Thyamine **T**	Phe	Ser	Tyr	Cys	T
	Phe	Ser	Tyr	Cys	C
	Leu	Ser	STOP	STOP	A
	Leu	Ser	STOP	Trp	G
Cytosine **C**	Leu	Pro	His	Arg	T
	Leu	Pro	His	Arg	C
	Leu	Pro	Gln	Arg	A
	Leu	Pro	Gln	Arg	G
Adenine **A**	Ile	Thr	Asn	Ser	T
	Ile	Thr	Asn	Ser	C
	Ile	Thr	Lys	Arg	A
	Met/START	Thr	Lys	Arg	G
Guanine **G**	Val	Ala	Asp	Gly	T
	Val	Ala	Asp	Gly	C
	Val	Ala	Glu	Gly	A
	Val	Ala	Glu	Gly	G

Note: Amino-acids are abbreviated as follows:

Phe – Phenylalanine
Leu – Leucine
Ile – Isoleucine
Met – Methionine
Val – Valine
Ser – Serine
Pro – Proline
Thr – Threonine
Ala – Alanine
Tyr – Tyrosine

His – Histidine
Gln – Glutamine
Asn – Asparagine
Lys – Lysine
Asp – Aspartic acid
Glu – Glutamic acid
Cys – Cysteine
Trp – Tryptophan
Arg – Arginine
Gly – Glycine

When a cell experiences a demand for a particular protein, an *enzyme complex* called *RNA polymerase* is guided to the segment of base sequences encoding that particular protein. It unzips the double helix and assembles a short single strand of a DNA-like molecule called *RNA* (*ribonucleic acid*) along the appropriate section of the DNA (Figure 3.8). By repeating this *transcription* process, multiple RNA copies of the relevant gene's base sequence are produced. This ensures that the *solitary* sequence of bases in the *long*

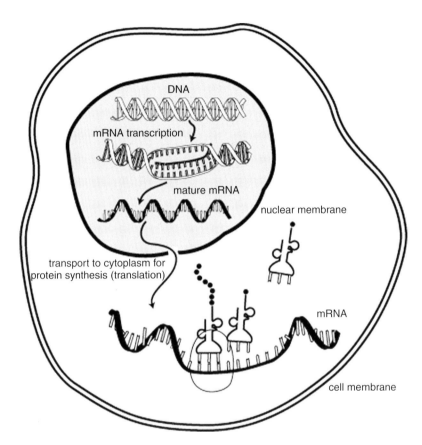

Figure 3.8 Messenger RNA is produced at the site of the appropriate gene in the nucleus and then transported to the ribosomes where the encoded amino-acids are joined together. Reproduced courtesy of National Institute of Health USA.

DNA molecule of the chromosome are expressed as *multiple* copies of *discrete* RNA molecules, each one of which carries the same information specifying the order of amino-acids in the encoded protein.

These multiple copies of so-called *messenger RNA* (mRNA) molecules move out of the nucleus to the ribosomes and, in doing so, convey the encoded information from the relevant sequence in the DNA to the 'shop floor' where the encoded proteins are assembled (see Figure 3.8). Ribosomes are robotic: given the appropriate RNA instructions, they can assemble any protein. At one time a given messenger RNA might encode the insulin protein; moments later another might specify an enzyme for utilising glucose. This process of building a protein based on the information encoded by an mRNA molecule is called *translation*.

The presence of this information retrieval system confers great flexibility on cells because regulatory factors which are active under certain conditions or in certain cell types determine which segment of DNA is available for unzipping (see Box 3.1). The mRNA, and of course its encoded protein,

Box 3.1 Regulating genes

All cells contain many thousands of specialised proteins called *transcription factors*. Their function is to interact in a very specific way with specific DNA sequences (called *promoters*) that are just before the DNA sequences that encode proteins. Some of these transcription factors are only active under certain conditions or in certain cell types; others are always active. They essentially determine which DNA sequences are unzipped and therefore ready for transcription. By regulating access to the genes these factors determine the production of proteins in the cells. The Figure 3.9 diagram below illustrates one of these interactions.

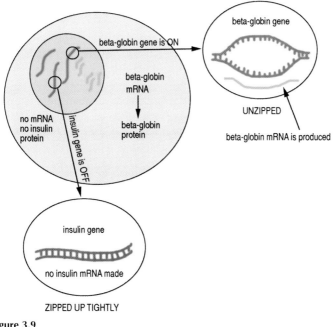

Figure 3.9

are therefore produced only when and where they are required. Thus, despite the fact that almost every cell in a human body contains *all* of the genetic information contained in the fertilised egg, different cells carry out different functions because *specific* sections of this information can be retrieved in different cells at different times. Thus, red blood cells contain thousands of messages encoding haemoglobin proteins but none at all for the insulin protein (not to mention thousands of others). In the appropriate cells in the pancreas, however, there are thousands of messenger RNA molecules for insulin but not a single haemoglobin messenger RNA molecule. The checks and balances ensuring the development of fertilised eggs into complex multi-cellular organisms like you and me are direct consequences of the remarkable level of gene regulation offered by these complex cellular processes.

Genes come in pairs

As mentioned earlier, human cells contain two complete sets of genetic information – one inherited from mother, the other from father. The information retrieval system uses the information from both for the assembly of required proteins. If a DNA sequence that codes for a functional protein has been inherited from both parents then 'normal' proteins are made in the appropriate cells (Figure 3.10 (a)). If a DNA sequence coding for a dysfunctional protein has been inherited from both parents then only abnormal proteins can be made. The cells containing these abnormal proteins will be compromised and such an individual will almost certainly manifest a pathological condition (Figure 3.10 (b)).

A DNA sequence coding for a functional protein is referred to as a 'normal' gene; a coding for a dysfunctional protein is a 'mutant'. If an individual inherits a normal gene from one parent and a mutant gene from the other parent, the information retrieval system will generate normal and abnormal versions of the same protein. Depending on the protein involved, these can have a variety of interactions resulting in a spectrum of effects for an individual who has inherited the two genes (Figure 3.10(c)). Mutated genes are referred to as:

- *recessive* if, under these circumstances, their presence fails to give rise to an observable condition;
- *co-dominant* if the individual exhibits an intermediate form of the disease; and
- *dominant* if, despite the presence of the other normal genetic information, the single mutant gene gives rise to the onset of the full-blown genetically inherited condition.

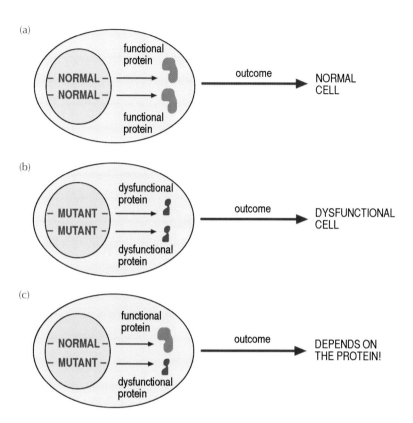

Figure 3.10 (a) Cells containing two copies of the 'normal' gene produce functional protein.
(b) Cells containing two copies of the 'mutant' gene produce dysfunctional protein which frequently manifests as a cellular pathology.
(c) Cells containing one copy of each produce both functional and dysfunctional protein which can produce normal, partially affected or abnormal cells depending on the protein.

Mutations and defective proteins

A comparison of the base sequence of the gene responsible for the assembly of haemoglobin from an individual suffering from sickle cell anaemia with the same gene segment from an unaffected individual reveals a very minor change in the DNA sequence. Just 17 bases into the gene sequence there is an 'A' in the normal gene and a 'T' in the mutant one. The decoder (Table 3.1) reveals what that means in terms of amino-acids: the first five amino-acids are the same for both proteins but the glutamic acid in the normal protein is replaced by a valine in the abnormal one. This alteration of one base in the DNA changes the amino-acid in position 6 of this protein. This particular protein is 146 amino-acids long so there are $3 \times 146 = 438$ bases to encode it; just one incorrect base is enough to cause this disastrous alteration in the amino-acid sequence and all the medical and personal consequences flowing from it (Box 3.2).

The story is similar for the defective protein associated with cystic fibrosis mentioned in Chapter 2. This protein is large: it normally has 1,480 amino-acids and therefore a coding sequence 4,440 bases long. A comparison of the DNA sequences of this protein from a cystic fibrosis patient with the same gene segment from an unaffected individual reveals that the first 1,521 bases are identical for both DNA sequences. The next triplet (TTT) which is found in the normal gene is completely absent from the mutant one but, thereafter, the other 2,915 bases are the same. This deletion of just one triplet in the mutant DNA means that the amino-acid which should occupy position 508 (Phe from the decoder) is simply not present in the mutant

---

Box 3.2 Mutations and mutations

Mutations vary in their seriousness. Some are so trivial that they do not change any amino-acids at all; that is because most amino-acids have more than one codon, so a mutation which changes one of the codons for a particular amino-acid for *another for the same amino-acid* has no consequential effect; it is called a "silent mutation". You can see from Table 3.1 that the amino-acid leucine, for example, can be encoded by CTT or CTC (as well as several others). Changing that final T into C is still going to specify leucine.

Some mutations might indeed cause a change in the amino-acid sequence but, because the two amino-acids exchanged are so similar in structure and function (say, leucine and isoleucine, or isoleucine and valine), there is hardly any effect on the way the protein works so there are essentially no consequences — the likelihood is that it would never be noticed. In theory, at any rate, some proteins might actually improve a protein's function but that is very improbable. The structures of today's proteins have been honed through selecting the best-performing versions resulting after billions of years of mutations; there is little chance that yet another one tomorrow will make a protein function better but every chance that it will make it worse. Indeed, the vast majority of mutations compromise function of a protein by distorting its structure so that it no longer works properly or even at all.

protein. Despite the presence of the other 1,479 amino-acids, this loss causes a subtle change in the shape of the protein which precludes its entry into the cell membrane. The cells in the body that need that protein in order to work properly are therefore compromised and the entire organism suffers as a result of this dysfunction.

Inherited misinformation

Cystic fibrosis is an example of a recessively inherited trait. The parents of a child suffering from cystic fibrosis have no symptoms of the disease – yet clearly they must have possessed genetic information producing a dysfunctional protein when passed on to their offspring. In this case each of the parents has one mutated and one unmutated instruction in their cells. Although able to produce both versions of the protein, their cells function normally because the abnormal protein does not interfere with the functioning of the normal one and sufficient quantities of the latter are made. This allows the parents to remain oblivious to the fact that they carry the dysfunctional information.

Such individuals will pass on *one* of these instructions to their offspring. If by chance an egg containing the mutated gene is fertilised by a sperm whose gene is also mutated, the resulting individual will have two copies of the mutation and thus have the instructions for making the abnormal protein only. That person will therefore develop the disease. It is just as likely that the normal gene is present in both egg and sperm. Such an individual inherits two copies of the genetic information for the assembly of the normal protein and will not therefore develop the disease. Finally, if one parent transmits a mutant gene with the other conferring a normal one, the resulting individual will have the same combinations of these genes as their parents: they will *carry* the information for the disease, and can pass it on to the next generation, but will not themselves become sufferers (Figure 3.11 (a)).

In such families, 1 in 4 of the children will suffer from the disease, 1 in 4 will be entirely normal and 2 in 4 will be carriers but not themselves affected. This is a totally random process and explains why individuals need not manifest a genetically inherited condition even when one or more siblings suffer from the disease.

Sickle cell anaemia is an example of a genetically inherited condition in which the normal and abnormal proteins can interact in a co-dominant fashion. As is the case with cystic fibrosis, individuals who inherit the gene for the dysfunctional haemoglobin protein from both parents develop the disease, whereas those inheriting the unmutated form of the gene from both parents do not.

Individuals who inherit a mutant gene from one parent and a normal gene from the other have red blood cells that contain both forms of the protein. When these red cells have a lot of oxygen, the abnormal protein does not make its presence felt and the cells retain their normal shape. Despite the presence of the *normal* protein, when such cells are starved of oxygen for a prolonged period of time the abnormal protein forms fibres, with a consequent alteration in the shape of the cells. People with this condition are described as suffering from *sickle cell trait*; they become unwell for a while until the body removes all of the dysfunctional cells and the attack of sickling of the red blood cells subsides again. Thus, unlike *carriers* of cystic fibrosis who possess a mutated version of the gene and can pass it on to the

(a)

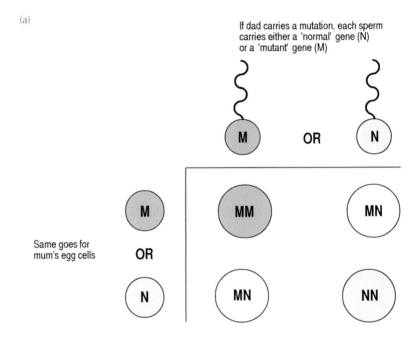

Dominant mutations are powerful because
only ONE parent has to have a mutant gene

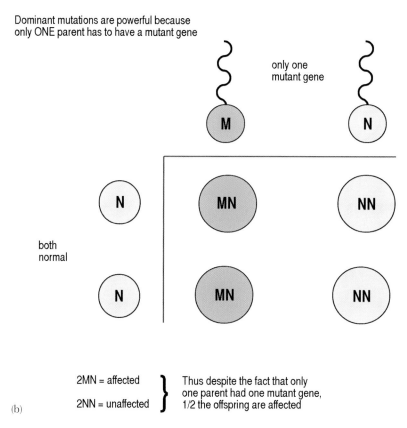

only one
mutant gene

both
normal

2MN = affected
2NN = unaffected

} Thus despite the fact that only
one parent had one mutant gene,
1/2 the offspring are affected

(b)

Figure 3.11 (a and b) Mutant genes differ in their ability to affect those that inherit them.

next generation without themselves suffering from the consequence of the abnormal protein they make, individuals *carrying* a mutated version of the sickle cell anaemia gene are subjected to bouts of a mild form of the same disease. Disease effects in families like this are 1 in 4 offspring showing disease symptoms, 1 in 4 unaffected and 2 in 4 exhibiting sickle cell trait (Figure 3.11 (a)).

Recessively inherited traits show just how important it is to have two complete sets of genetic instructions per individual. Every single fertilised egg contains a number of seriously mutated genes – some are inherited from the father, others from the mother. As long as one parent provides an unmutated version, the individual will normally be oblivious to the mutation carried because the transcripts from the normal gene are often sufficient to provide an acceptable level of normal protein. It is only relatively rarely, when an individual inherits similarly mutated genes from both parents, that normal protein is totally absent and the disease is evident.

However, in other, and thankfully rarer, cases the inherited dysfunction *dominates* the normal function: this means that the presence of one copy of the incorrect information is sufficient for the full disease to appear. Huntington's chorea is just one example of this type of inherited disease. The inherited misinformation causes the production of an abnormal protein in the nerve cells in the brain. The abnormal protein interferes with the normal one so that, even though the cells are still producing lots of the normal protein, they cannot carry out their function and nerve degeneration becomes evident in late middle age. This manifests as early senile dementia with total mental and physical disability 10–20 years after the onset of the initial symptoms. As this is a dominant mutation, individuals who inherit just *one* copy develop the disease. But even more tragically, by the time they realise that they have the gene they have already had their own families and passed on their genes to the next generation. Regardless of the genetic information from the other parent, on average, half their offspring will by then have inherited the same dysfunction and will be destined to develop the disease (Figure 3.11 (b)).

Summary

We have seen that a gene is a particular combination of bases (A, T, G, C) in one section of the long DNA molecule. This information is decoded by the cell's sophisticated gene reading and decoding devices to produce a string of amino-acids joined together in a predetermined order as a protein. A change in such a DNA is referred to as a mutation. Some mutations have no apparent effect on the protein but many result in a dysfunctional protein.

Every fertilised egg contains a number of mutated genes – some are inherited from the father, others from the mother. As long as one parent provides an unmutated version of the gene, the presence of one normal gene is often enough to ensure an acceptable level of normal protein and the individual may be unaware of the mutation. This is the case in cystic fibrosis: one copy of correct information provides a sufficient number of functional pores to allow cells to function normally. In sickle cell anaemia, the presence of high levels of functional haemoglobin is sufficient to prevent cellular dysfunction, except under the physiological stressful conditions of oxygen deficit.

If, however, an individual inherits similarly mutated genes from both parents, the complete lack of normal protein usually precipitates a pathological condition. Conditions which come to light only when defective genes are inherited from both parents are recessive genetic traits. In dominantly inherited traits, the mutated gene codes for an altered amino-acid sequence with a new and deleterious function that the normal protein, which the individual also possesses, cannot mask.

The inherited conditions on which we have focused thus far arise from defects in a single gene and give rise to reasonably straightforward inheritance patterns. Human genetics is rarely this simple, however, and after 25 years of intense research we still find that the more we know, the more we know we don't know!

The health of the nation

It does not require training in genetics to realise that many common ailments run in families. High blood pressure, heart attacks, diabetes, mental disorders, autoimmune diseases (such as arthritis) and cancer account for the majority of the debilitating and often fatal diseases afflicting those living in the Western world. Unlike the diseases we discussed in the last chapter, in which one malfunctioning protein causes a serious dysfunction, no *single* defect gives rise to these common diseases. Instead a *number of different* genes are involved in determining a person's predisposition to each of them. When five different proteins are involved in a cellular function (and normally there are many more than that), there are over 200 different ways the genes that encode them can be inherited (see Box 4.1). Small wonder then that the predictive power of genetics that is so useful in single gene trait inheritance is next to useless here. Furthermore, environmental factors such as diet, lifestyle, family environment and exposure to pollution play an enormous role in determining whether or not an individual will develop these diseases. Despite the daunting prospect posed by such complexity, researchers have already made a start at disentangling the contributions to these conditions made by different genes.

Affairs of the heart and head

Heart attacks and strokes are the direct consequence of an interrupted blood supply to the heart muscle and brain, respectively, caused in turn by occluded arteries. The two major risk factors for strokes and heart attacks are high blood pressure and a high level of cholesterol in the bloodstream. The body uses an array of different proteins to regulate these two complex systems and therefore their genetic inheritance patterns are very complicated. Analysis of the genes regulating the four major proteins responsible for overall control of blood pressure has revealed a number of interesting observations, but a real understanding of the system requires the investigation of yet more proteins and the genes that encode them. The genetic factors controlling blood pressure are complicated enough but, when compared to those

Box 4.1 The inheritance patterns of complex traits are obscured by multiple gene combinations.

We have previously seen that there are three possible patterns of inheritance when **a trait** is determined by **a single gene** encoding the structure of **a vital protein** such as hemoglobin (see Figure 3.10). This is because we carry two copies of the genetic material in our cells (one copy we inherited form our father; the other from our mother). A person can have two dominant forms of the gene (represented as AA), two recessive forms of the gene (represented as aa), or one of each (represented as Aa).

   Different genes encode **different proteins** and can be named B, C, D etc to designate that these sequences are found at separate locations along the DNA. Each on of these can be dominant or recessive giving BB, bb, Bb; CC, cc, Cc; DD, dd, Dd etc.

   If **one trait** is controlled by **two *different* genes**, located at positions A and B respectively along the DNA, there are nine (3×3) different possible combinations. Each additional gene multiplies these combinations by a further factor of three.

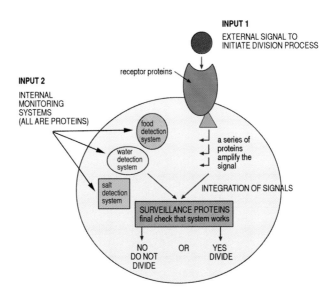

Figure 4.1 An intricate interaction of hundreds of different proteins is required every time one cell divides into two.

regulating cholesterol levels, they are positively clear-cut. Nothing, it seems, is simple in affairs of the heart.

Cholesterol is an essential constituent of our cells, normally absorbed from the intestine or made in the liver. It has to be transported around the body but it does not dissolve in the watery part of our blood, so the body encases the cholesterol to facilitate its transport through the bloodstream. Special proteins embedded in these particles enable them to deliver their valuable contents by docking onto receptor proteins in the cell membranes where it is needed. At the last count there were more than a dozen different genes encoding proteins involved in the complex production, transport and release of cholesterol around the body. If any of them is dysfunctional, the system becomes inefficient and cholesterol starts to accumulate in the bloodstream where it gets deposited on the walls of the blood vessels, thus starting the gradual occlusion of arteries to vital organs. With the exception of a rare recessively inherited mutation which eliminates the receptor protein from their cells, attempts to understand the genetic basis of strokes and heart attacks are still in their infancy.

There is absolutely no doubt that mutations in some of the genes encoding these proteins can cause serious ill health, and the major ones responsible for a predisposition to these diseases will no doubt eventually be identified. Until then, studies reveal that for reasons not yet entirely clear, a couple of glasses of red wine every day is an effective way of reducing the chances of suffering a heart attack. Furthermore, while low cholesterol diets improve your chances of avoiding strokes and heart disease, the lack of this essential ingredient in the body may give rise to other complications (including a deterioration of mental skills due to nerve dysfunction). Since there is no convincing evidence that low cholesterol diets increase longevity, it would therefore appear to be safe to eat a nutritious meal with the aforementioned wine.

Cancer

Cancer describes a condition of uncontrolled cell division in a particular part of the body. Although 'war' was declared on this killer over 30 years ago, the word still strikes fear into our collective consciousness. The reason is that the simplicity of the description belies the fact that cancer is neither a simple target nor indeed even a single disease. That is why, despite the fact that hardly a year goes by without a claim of a new breakthrough in our fight against the disease, the war goes on.

An intricate interaction of hundreds of different proteins is required every time one cell divides into two (Figure 4.1). Some proteins signal the cell to start dividing, many more are involved in the process of division itself, whereas others are responsible for ending the process when the two new cells have been formed. Cancer arises when one or more of these proteins becomes defective and this balance is disturbed – cells either start to divide when they should not or continue to divide after they should have stopped. With so many proteins involved in the process, small wonder that there are more than 200 different ways in which the human body can experience uncontrolled cell division – each one of which gives rise to a different form of cancer.

To develop a deep understanding of even one of these proteins is difficult; to discover how it exerts an effect on the development of even one type of cancer is even more demanding; and to use the information to evolve an effective therapy is nothing short of astounding. A lot of work remains to be done before the multi-faceted origins of this complex of diseases are properly understood, but many wonderful breakthroughs have nevertheless been made in the prevention, diagnosis and treatment of cancer – and many more will undoubtedly be made in the years to come.

Cancer arises due to defective proteins, defective proteins arise due to gene mutations, yet cancer is *not* an inherited disease. Although a predisposition to some forms of cancer can be inherited, the majority of the mutations responsible for the development of uncontrollable cell proliferation occur as our bodies mature and grow old. Every time a cell divides, its DNA is replicated. The enzymes responsible for synthesising the new DNA molecules make occasional errors – about one base per million incorrectly copied – and so mutations accumulate as cells divide. Moreover, environmental factors may accelerate this process because they can cause damage to the DNA of dividing and non-dividing cells alike. Mutations arising in the body cells are restricted to the individual in which they arise and are therefore not passed on to the next generation.

Factors damaging DNA (and thus inducing mutations in the genes controlling cell division) include tobacco smoke, radiation (e.g. ultraviolet light from the sun), X-rays and radiation from radioactive materials, various chemicals, asbestos, a number of viruses and certain types of environmental pollution. These mutations do not necessarily cause cancer immediately because nature uses a 'belt and braces' approach to this most critical of cellular functions, but they make its occurrence much more likely. Whether or not tumour formation actually takes place comes down to a combination of:

- the genes we inherit from our parents,
- the environment we live in,
- the number and type of mutations that we accumulate, and
- good old-fashioned luck.

The catastrophic cascade to a tumour

The single fertilised egg with which each one of us commenced our existence contained all the genetic information required to produce a mature human being. Cell growth and division was central to this task and every cell containing a nucleus carried every one of the genes necessary for this process. The stimuli for a cell to divide include:

- external signals (e.g. growth factors secreted from one type of cell which bind to a protein receptor in another) telling it to divide;
- internal signals (the nutritional status of a cell is constantly being monitored by other proteins to assess whether it contains sufficient food to divide) telling it that it can do so if it has been so directed;
- the appropriate interaction of these two systems with one another and other systems within the cell.

At the last count, more than 70 genes had been identified as being intimately involved in the mechanism of cell division control. Each one of these genes is essential for the development and maintenance of the body and, when their encoded proteins function correctly, they ensure that cells divide only where and when they are required to do so. Should any one of these acquire a mutation (due to DNA damage) resulting in the production of a defective protein, the affected cell attempts to divide in an unregulated fashion. Such mutated genes are given the generic name 'oncogene' because they give cells a signal to divide in an unregulated fashion. However, cells also possess a series of other genes which encode a sophisticated surveillance system to 'audit' many factors before division can proceed. These so-called *tumour suppressor genes* produce proteins that prevent mutated cells from dividing uncontrollably (Figure 4.1), a fail-safe mechanism preventing the first mutation from causing havoc. However, should a second mutation occur in one of the other genes, the tumour suppressor system becomes ineffective and rapid cell division frequently commences.

Even after the control systems have become severely compromised and the cells are dividing rapidly, they are not completely out of control. The accumulating runaway cells continue to divide until they undergo a crisis during which many of them die. The survivors emerge with gross chromosomal alterations, some of which have inactivated other control genes, giving rise to an uncontrolled cell division which rapidly develops into a tumour. The bottom line is that the conversion of a normal cell to a tumour cell is not a single event but requires a number of separate steps occurring over a period of time (Figure 4.2).

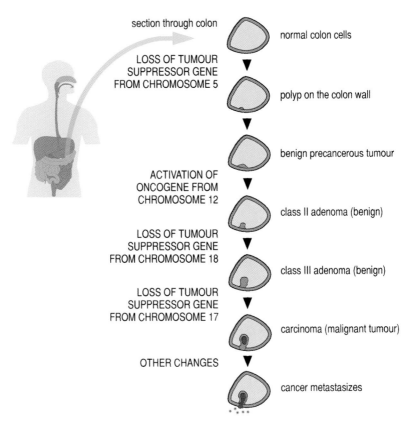

section through colon

normal colon cells

LOSS OF TUMOUR
SUPPRESSOR GENE
FROM CHROMOSOME 5

polyp on the colon wall

benign precancerous tumour

ACTIVATION OF
ONCOGENE FROM
CHROMOSOME 12

class II adenoma (benign)

LOSS OF TUMOUR
SUPPRESSOR GENE
FROM CHROMOSOME 18

class III adenoma (benign)

LOSS OF TUMOUR
SUPPRESSOR GENE
FROM CHROMOSOME 17

carcinoma (malignant tumour)

OTHER CHANGES

cancer metastasizes

Figure 4.2 The conversion of a normal cell into a tumour is not a single event.

Inherited predisposition to cancer

Except for the odd exception (which biology always generates to keep us guessing), the copies of the 70 genes so far identified telling a cell to divide – genes we inherit from our parents – are normal. Mutations accumulate only during our own individual lifetimes. The longer we live, the more chance there is of accumulating a mutation which is one reason why cancer is predominantly a disease of the over 60s. But no age group is immune from this disease; many factors contribute to tumours, some of them are inherited, others not.

Inherited mutations of tumour suppressor genes
We have just seen that tumour suppressor genes produce proteins to counter-

act the influence of dysfunctional cell control proteins. However, it is possible to inherit a mutated version of one of those genes. This means that the fail-safe is compromised and, *should* a mutation occur in any one of the 70 or so genes controlling cell division, the cell is two steps (rather than one) along the route to a tumour. This is how a predisposition to developing cancer can be inherited. As long as the cell division genes remain unmutated, the compromised fail-safe mechanism goes undetected, but the inherited mutation is nevertheless a ticking time bomb awaiting DNA damage in some of the other genes. Individuals with such a mutation tend to acquire cancer at a much earlier age than would normally be the case.

Inherited deficiency in DNA repair

Every cell containing DNA also has a set of enzymes which recognise and repair damage in DNA, whether arising during cell division or because of an environmental effect. However, such proteins are also encoded by genes. Should an individual inherit genes encoding defective DNA repair proteins, his cells will not be able to repair DNA as efficiently as is normally the case. Unsurprisingly, such individuals accumulate lots of mutations in their cells, some of which lead to the cell control genes being deactivated; the affected individuals are therefore strongly predisposed to developing cancer.

Environmental influences

Viruses and tumour formation

The small growths we call 'warts' are in fact benign tumours causing no lasting difficulties. Any one of 60 different *papilloma viruses* can cause warts but only a few confer a predisposition to cancer. Around 80% of all cervical cancers show evidence of infection by just two particular types of these viruses; the additional factors needed to convert viral infection to cancer are unknown. Nevertheless, a vaccine aimed at preventing infection by the two viruses is currently undergoing clinical trials with a view to preventing infection and thereby eliminating the major predisposing factor for this disease.

Tissues affected by reproductive hormones (e.g. cervix, breasts, testes)

Such tissues are prone to cancer in people of reproductive age. Cervical cancer is a good example of normal cells taking time to become tumours; pre-malignant cells can therefore be detected in smear tests (Figure 4.3) and eliminated before they cause trouble. Likewise, self-examination for abnormal lumps in the breasts or testicles can catch the disease in its early stages.

Environmental mutagens

We have seen how cancer normally requires mutations to occur in two or

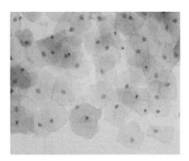

Figure 4.3 Cervical smears catch cells before they turn malignant. Reproduced courtesy of Medical School QMUL.

more genes concerned with the regulation of cell division. It therefore follows that any factors increasing the frequency of mutations will also increase the tendency for cancer cells to arise; conversely, anything decreasing mutation frequency delays cancer cell formation. This consideration underlies concern about excessive exposure to bright sunlight and the depletion of the ozone layer resulting in more skin cancer. The ozone layer normally absorbs ultraviolet light from the sun. In its absence, these rays in sunlight cause DNA damage much more often than the repair system can accommodate. As a result, mutations are generated, some of which induce oncogenes and subsequently trigger the development of a skin cancer.

Tobacco smoke
Tobacco smoke is an even more potent source of mutagens. There are over 6,800 different chemicals in tobacco smoke and several of them cause mutations. Thus, the prolonged exposure of lung tissue to smoke causes the accumulation of oncogene mutations and the eventual onset of lung cancer. Tobacco smoke is so powerful that even the urine of smokers contains mutagens (Figure 4.4); that is why smokers suffer not only from lung tumours but cancer of the bladder as well. In brief, a smoker is 1,000 times more likely than a non-smoker to get cancer and it is estimated that, of smokers who die from lung cancer, 85% would not have done so if they had not smoked. Lung cancer is one of the biggest killers in the Western world. It has little to do with genetics but everything to do with inhaling toxic fumes from cigarettes.

However, not all environmental influences are malign. Fresh fruit and vegetables contain compounds which prevent DNA damage. Thus, individuals can dramatically decrease the possibility of developing a whole

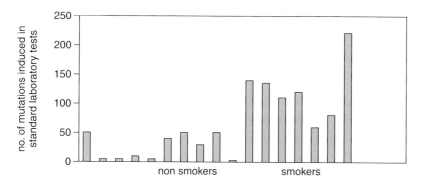

Figure 4.4 Even the urine of smokers contain chemicals that damage DNA.

range of cancers by adopting a healthy diet and practising some common sense with regard to sunbathing – but overwhelmingly by avoiding exposure to cigarette smoke.

Genes and the brain

If, because of the complex interactions between scores of proteins and environmental factors, genetics has difficulty in predicting the appearance of cancer, the inheritance patterns of brain function/dysfunction are even more impenetrable. It is conservatively estimated that more than 20,000 genes are involved in the development and maintenance of the human brain and that there are 100,000,000,000 individual brain cells. Brains are complicated, the environment in which we are reared determines our personality to a large extent and many mental disorders are difficult to categorise accurately. It is safe to say, therefore, that any serious understanding of genes and brains, not to mention genes and minds, is still a long way off.

Some mental disorders, like Alzheimer's disease (senile dementia) and Creutzfelt-Jacob's Disease (CJD), are associated with obvious changes in the physical structure of the brain. As nerve degeneration is accompanied by changes in proteins, and as genes determine the latter, an analysis of genetic predisposition to these conditions promises to reveal much about the molecular basis for these diseases. Alzheimer's disease has already become one of the health problems of the developed world. By the age of 65, 1% of the population has it; by the age of 85, it is closer to 20% (Figure 4.5). In some cases it is clearly inherited, in others the genetic basis is a lot less clear. In the lineages of some patients, a number of mutations in a particular gene encoding the *β-amyloid protein* (which behaves in a defective way in the

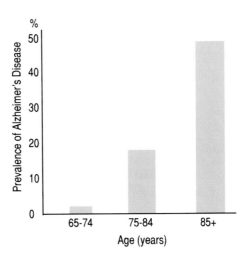

Figure 4.5 The incidence of Alzheimer's disease increases with age.

nerves of Alzheimer's sufferers) has been associated with this complex disease but other families who inherit the disease have normal copies of the gene. Instead, another gene has been implicated, one which encodes a protein helping to transport cholesterol around the body and interacting with the β-amyloid nerve protein. As time passes, more and more candidate genes are being identified, but even this extensively studied brain dysfunction still requires a proper explanation.

The story for Creuztfeld-Jacob's Disease (CJD) is similar. This condition is characterised by progressive nerve degeneration leading to physical incapacity, dementia and death. At autopsy, the sufferer's brain has characteristic 'holes' in the tissue. There are two forms of this disease: an *inherited* version and a *sporadic* type which appears in individuals without a history of CJD in their family. Genetic analysis of individuals with the inherited form of this disease has identified a gene for a protein called *PrPC* that is present in nerve cells. When certain mutations in this gene are inherited, the encoded protein has an abnormal shape and, over time, causes so much cell disruption that nerves start to die off leaving those holes in the brain. The progressive nature of the disease is due to this increasing brain degeneration.

The impact of genes on the sporadic form is not quite so clear-cut. Individuals who get the disease frequently have genes encoding slightly different versions of the PrPC protein from those individuals in the population who have not developed the disease. However, many believe that an infectious protein (a *prion*) is responsible for this form of CJD. The acquisition of the latter has been associated in some areas with eating undercooked sheep's eyes and brains, and in others with exposure to contamination by human nerve tissues such as that from corneal tissue grafts or instruments used in neurosurgery – and with the ritualistic eating of human brains. *New variant CJD*, a particularly topical example of the sporadic form of this disease, has been closely linked with the internationally infamous BSE (*bovine spongiform encephalopathy*; mad cow disease) outbreak in Britain. In this case, the source of the problem appears to be beef from cattle which have been fed contaminated sheep brains. It has been suggested that, whatever the route of infection, prions enter the body and then make their way to the brain where they interact with the PrPC protein normally found there, causing it to change its shape. Then, as appears to be the case with the inherited form of the disease, a chain reaction of protein deformation, nerve death and brain degeneration results. It is likely that certain forms of the protein are more susceptible to this phenomenon than others, and it is therefore possible that new variant CJD may be limited to persons who have the susceptible gene(s) *and* have also been exposed to infection by the infectious prion particles. (Note that not all scientists accept the prion theory of BSE and CJD, so a final agreed explanation must await further exploration.)

Alzheimer's disease and CJD are associated with obvious changes in the physical structure of the brain. With other dysfunctions, such as manic-depression and schizophrenia, there are no obvious changes in the physical structure of the brain, yet the conditions are treatable with appropriate drugs; this suggests that they arise because of imbalances in the brain's chemical messengers, which again indicates at least a strong genetic component to the disease. A number of genes encoding brain proteins which transmit signals from one cell to another have been analysed (Chapter 5 will explain how this is done) and a few genes have been associated with a predisposition to psychological problems. However, the isolation of major genes involved in serious conditions like manic-depression and schizophrenia is still some way in the future.

Finally, whereas we possess some evidence of a genetic basis for brain diseases and psychological dysfunction, elucidating the genetic foundation of an individual's complex behavioural patterns becomes increasingly elusive as the level of complexity increases. Behavioural differences between individuals are the result of multifaceted interactions between the proteins encoded by their genes, their social background and their physical environment. Such higher-order behavioural patterns include:

• talents or inabilities for various activities – music, sport, academic pursuits;

- sexual orientation;
- intelligence;
- a person's overall disposition and the myriad traits which make each individual unique.

As of now, despite reports in the press about the discovery of a *gay gene* or *intelligence genes*, claims to have discovered 'the genetic basis' of higher-order behavioural activities have more to do with hype than hard scientific facts. No doubt genes do play a role in providing a predisposition to many behavioural traits, but the subtle interactions of other brain centres and proteins, our environment and even the food we eat can have far-reaching effects on the integration of our behavioural patterns. There is a lot more to brains than just the genes.

In conclusion

Many of the genes we carry are common to all of us as humans but each one of us also carries a unique contribution of minor variations in the form of dominant, co-dominant and recessive forms of many genes; these have the cumulative effect of making each one of us unique. Some, like those affecting height, the colour of eyes, hair and skin, or blood group have little or no effect on the health of an individual. Others which govern cholesterol metabolism, tumour suppression and the PrPC protein have profound influences. Many human diseases are associated with defective proteins, the information for which is to be found somewhere within that individual's DNA. Some mutant genes are inherited while others arise during the lifetime of the individual. Either way this leads to two unavoidable conclusions:

- if the language of the genes can be understood it is easier to understand the disease itself;
- if this information can be appropriately manipulated, there is a better chance that the disease can be cured.

To accomplish these objectives, it is necessary first to isolate the offending gene and produce many millions of copies to facilitate a full analysis. This is called *gene cloning* and we now go into a genetics laboratory to find out how it is done.

5

Dealing with the invisible

atggtgcacctgactcctgaggagaagtctgccgttactgccctgtggggcaaggtgaacgtggatgaagttgg
tggtgaggccctgggcaggctgctggtggtctacccttggacccagaggttctttgagtcctttggggatctgtcc
actcctgatgctgttatgggcaaccctaaggtgaaggctcatggcaagaaagtgctcggtgcctttagtgatggc
ctggctcacctggacaacctcaagggcacctttgccacactgagtgagctgcactgtgacaagctgcacgtg
gatcctgagaacttcaggctcctgggcaacgtgctggtctgtgtgctggcccatcactttggcaaagaattcacc
ccaccagtgcaggctgcctatcagaaagtggtggctgggtggctaatgccctggcccacaagtatcactagct
cgctttcttgctgtccaatttctattaaaggttcctttgttccctaagtccaactactaaactgggggatattatgaag
ggccttgagcatctggattctgcct

Asked to scan through the text in this book to pick out the section containing the genetic code for part of the human globin gene, few people would have difficulty guessing that it must be the line of letters just below this chapter title. It looks different from everything else! Suppose, however, that *every* page of text was a continuous string of As, Ts, Gs and Cs and if, instead of being presented with 100 pages of text, ten million pages were offered for examination, most searchers would simply throw up their hands in horror. Now imagine that the text had to be found without the reader being able to see the letters at all; then you begin to get a sense of what it is like to be a hunter of human genes! These unfortunate scientists have first to separate their gene of interest from all the other three billion bases in the DNA of each human cell and then find some way of visualising the order of its bases to reveal the information it encodes. Small wonder that they hijack Mother Nature's little tricks in their quest.

Doing what comes naturally

Bacteria are ubiquitous, microscopically small, single-celled organisms, far too small to be seen without a microscope. Simple in structure, they reproduce by dividing in two to generate two identical cells. Some can divide every 20 minutes or so and thus, if sufficient nutrients are available (the record is about 9 minutes), can produce many billions of progeny overnight. If a number of bacteria are spread onto the surface of a nutrient jelly providing every food component they need to grow, each tiny bacterial cell will divide and divide to produce in a few hours a jumbled heap of cells in a *colony* which can be seen with the naked eye. The glass or plastic vessels used to grow bacteria are called *Petri dishes* after their inventor; microbiologists usually refer to them as 'plates', so spreading bacteria onto nutrient jelly is known as 'plating'.

Bacteria need considerably fewer genes than complex human cells in order to survive; in fact, all 3,000 genes that they possess are carried on a single circular chromosome. Some bacteria possess a second, much smaller circular chromosome consisting of genes that are useful but not essential for survival. These circular DNA molecules are called *plasmids*; they vary in size, type and the number of genes they carry (Figure 5.1).

When a bacterial cell divides, or an organism produces offspring, genetic information is normally transferred *vertically* from one generation to the next. Many plasmids, however, have a series of genes encoding the formation of a conjugation tube to facilitate mating and the transfer of genes from one bacterial cell to another. This constitutes an exchange of genetic information between mature cells, an ability to transfer genes *horizontally* from one cell to another in the same generation (Figure 5.1). It is an extremely useful property for the bacteria because it allows a large population of cells to acquire new traits very rapidly. Such plasmids often carry genes coding for resistance to antibiotics. The ready transfer of bacterial plasmids, however, is not at all useful for us: it is largely responsible for the rapid appearance of multiply-resistant bacteria in hospitals because of the ease with which several different antibiotic resistance genes are able readily to be transferred from one bacterial cell to another.

Decades ago microbiologists realised that if a particular plasmid were removed from the bacterial cells that carried it, just mixed with other bacteria in a test-tube and the cells given a brief heat shock, a small proportion of them absorbed the naked plasmid DNA. Each of these cells picked up just one plasmid but in doing so acquired the gene that made them resistant to the antibiotic (Figure 5.2). Spreading the absorbing cell culture onto a plate of jelly containing the food they needed for growth, together with the antibiotic, identified those cells that had picked up the resistance gene. This was because the antibiotic killed off cells that had not acquired the genetic information; those that had grew into visible colonies overnight (Figure 5.3). The plasmid also replicated at every cell division generating many million copies of itself by the morning.

Figure 5.1 Plasmids facilitate mating and the exchange of genetic information between bacterial cells.

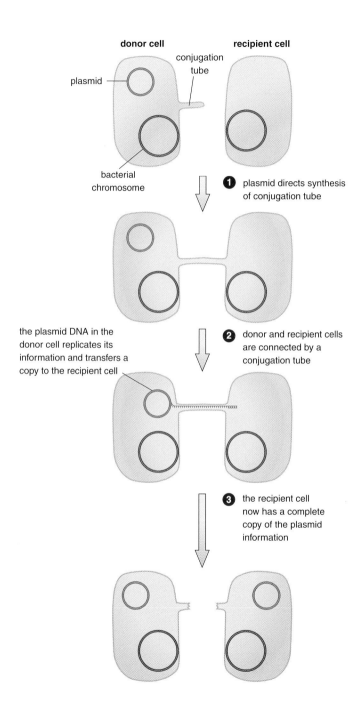

donor cell   **recipient cell**

plasmid

conjugation tube

bacterial chromosome

1. plasmid directs synthesis of conjugation tube

2. donor and recipient cells are connected by a conjugation tube

the plasmid DNA in the donor cell replicates its information and transfers a copy to the recipient cell

3. the recipient cell now has a complete copy of the plasmid information

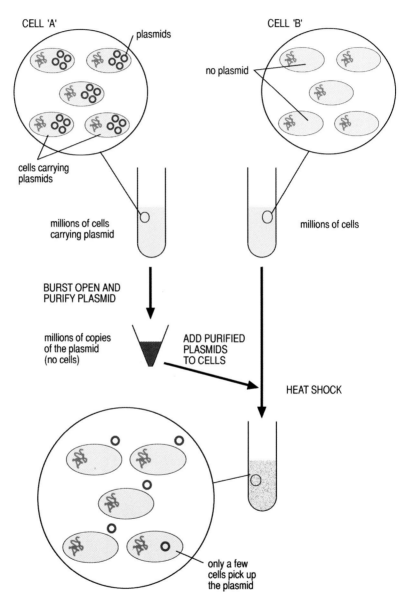

Figure 5.2 Scientists can 'persuade' bacteria to pick up plasmids directly from a test tube in a process called transformation.

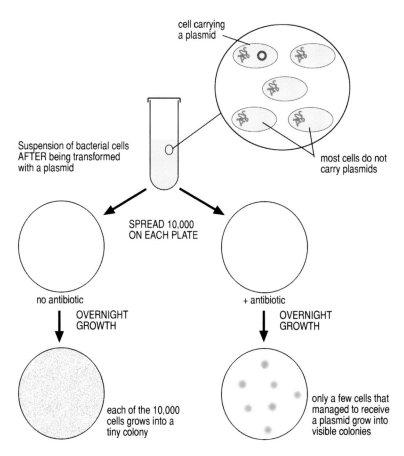

Figure 5.3 Only the rare cells that have picked up the plasmid from the test tube can grow into a colony on a nutrient medium containing the appropriate antibiotic.

Hunters of human genes require many millions of copies of their gene of interest but of course human genes are on huge chromosomes inside their cells – not on plasmids inside bacteria! However, if a specific section of human DNA could be tied into a plasmid in a test tube and 'smuggled' into a bacterium so that it replicated along with the plasmid, overnight, a gene hunter could obtain millions of copies of their gene of interest. Nice idea in theory but how might it be achieved in practice? The tools that gene hunters use for this sleight of molecule are all provided by Mother Nature herself. We have already met plasmids and bacterial cells; the remaining players are enzymes (proteins that make things happen).

All cut up

Bacteria are subject to attack by special viruses (called *bacteriophages*) which inject their own DNA into the bacterial cell and hijack it to make new viruses. Bacteria possess specialised enzymes that continuously scan *all* of the DNA in the bacterial cell; if they find a piece of viral DNA they slice it up in a very special way. Scientists can purify these so-called *restriction enzymes* and, fortuitously, if they are added to *any* DNA in a test-tube, they cut it with exquisite precision.

The first restriction enzyme was purified from the common bacterium *Escherichia coli (E. coli* among friends). The enzyme, termed called *EcoR1*, cuts DNA molecules wherever it finds the base sequence G-A-A-T-T-C in DNA regardless of the origin of the molecule. As this particular sequence will occur randomly in all long DNA molecules, the enzyme will cut the DNA into random fragments – about half a million of them in the case of human DNA. Because the enzyme is so specific, DNA is always cut into the same precise fragments whenever it is exposed to this particular enzyme. Further-more, the DNA is cut in a staggered fashion to produce short overhanging segments, all ending in CTTAA. So, after human DNA has been exposed to this enzyme, it is transformed from a series of incredibly long DNA mol-ecules of totally unknown DNA sequence into 500,000 smaller fragments. Each is still of unknown sequence but ends with the same short DNA handle on the right hand side and the precise *complementary* sequence on the left (Figure 5.4). Plasmids are, of course, made of DNA and are therefore sensi-tive to digestion by restriction enzymes. A circular plasmid cut with EcoR1 will become a linear piece of DNA *and carry exactly the same left and right handle sequences as every one of the 500,000 fragments of human DNA as prepared above.* The handles are very useful: they will eventually be exploited to smuggle each one of the individual fragments of human DNA into different bacterial cells.

Sealed with a kiss

Now comes the elegant step in this procedure. We have seen that in DNA molecules T is always joined to A, and G to C; when we add EcoR1-cut plasmid DNA to EcoR1-cut human DNA, the complementary 'handles' seek out one another and kiss gently together. Mixing them at 12°C prolongs the kiss because thermal agitation shakes them apart. Another enzyme, one that bacteria normally use for repairing their own DNA when it gets damaged

---

Figure 5.4 The restriction enzyme EcoR1 cuts DNA molecules, leaving a short over-hang, wherever the sequence GAATTC occurs.

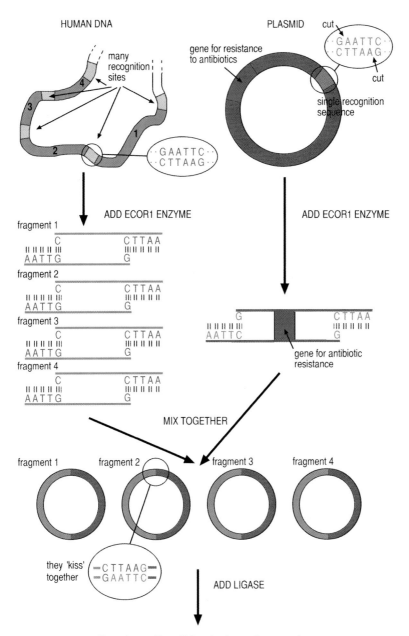

HUMAN DNA

many
recognition
sites

4

3

2

1

· G A A T T C ·
· C T T A A G ·

PLASMID

cut

gene for resistance
to antibiotics

· G A A T T C ·
· C T T A A G ·

cut

single recognition
sequence

ADD ECOR1 ENZYME

ADD ECOR1 ENZYME

fragment 1

C   C T T A A
AATTG   G

fragment 2

C   C T T A A
AATTG   G

fragment 3

C   C T T A A
AATTG   G

fragment 4

C   C T T A A
AATTG   G

G   C T T A A
AATTC   G

gene for antibiotic
resistance

MIX TOGETHER

fragment 1   fragment 2   fragment 3   fragment 4

they 'kiss'   C T T A A G
together   G A A T T C

ADD LIGASE

millions of recombinant DNA molecules, each one consists
of a plasmid containing a discrete segment of human DNA

inside the cell, can then be added to the cooled tube and the kissing molecules, as it were, are 'superglued' together. The two molecular fragments have become one *recombinant DNA* molecule. Next, the plasmids, each one with its newly incorporated *human* DNA fragment, are added to bacterial cells, heat shocked and placed on jelly containing the relevant antibiotic. Only cells that take up a plasmid can grow and the system can be biased to ensure that only those cells which have accepted a plasmid joined to human DNA fragment actually survive. Each cell carrying a plasmid will divide every 20 minutes, so providing a colony containing millions of cells by the next morning.

Think what that means in terms of copies. Let's say that a bacterium divides every 20 minutes. That means there will be two cells 20 minutes after we start, four cells after 40 minutes, eight after an hour, 64 after 2 hours, 512 after 3 ... more than 2 million cells in 7 hours and 4 trillion in 14! As there can be upwards of 20 copies of plasmid in each cell, overnight we can get 80,000,000,000,000 copies of the plasmid spliced together with the human DNA fragment. On the next day each colony on the plate will contain many millions of plasmids and, of course, millions of copies of the inserted individual human DNA fragment. Each colony thus contains a *clone* of a unique human DNA fragment.

One in a million

By plating enough cells we can produce in excess of one million discrete colonies and thus be pretty sure that the DNA sequence we need is present in at least one of them. We have now constructed a 'library' of fragments containing the information needed to assemble an entire human being. Instead of being contained in books, however, the information is encrypted in DNA molecules which have been spliced into plasmids carried in bacterial cells.

A special type of filter paper is pressed gently onto the plates of bacterial colonies so that a sample of cells from *each* colony gets transferred to the paper. The filter papers are then treated with a series of strong chemical reagents, first to break open the cells to release their DNA and then to cause the two strands of the DNA to separate, thus exposing their bases. All that remains is to find a way of identifying the position of the gene in which we are interested. Occasionally something is already known about the gene and that can be used to find it. The first human gene to be cloned like this from

Figure 5.5 The location of the colony bearing the human gene of interest can be visualised using a radioactive 'probe'.

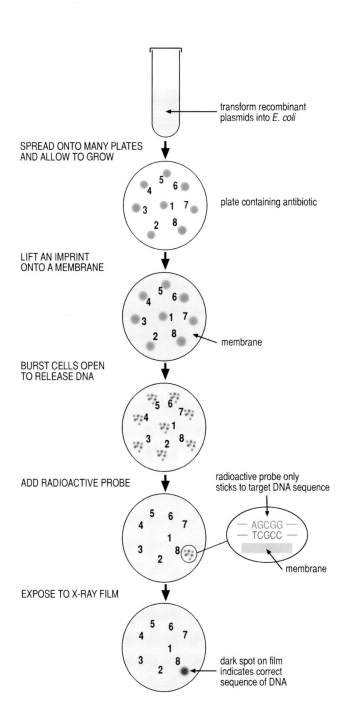

transform recombinant plasmids into *E. coli*

SPREAD ONTO MANY PLATES AND ALLOW TO GROW

plate containing antibiotic

LIFT AN IMPRINT ONTO A MEMBRANE

membrane

BURST CELLS OPEN TO RELEASE DNA

ADD RADIOACTIVE PROBE

radioactive probe only sticks to target DNA sequence

— AGCGG —
— TCGCC —

membrane

EXPOSE TO X-RAY FILM

dark spot on film indicates correct sequence of DNA

total human DNA was the one that encodes the protein involved in sickle cell anaemia, the beta-globin gene.

In this case, the identity of the offending protein was known and it was also known that red blood cells contain large quantities of it. Immature red blood cells contain millions of the messenger RNA molecules which carry the genetic information from the gene enabling the cells to make the protein. The messenger RNA was therefore removed from these red blood cells. A special enzyme called *reverse transcriptase* was added to the RNA and allowed to make one strand of DNA under conditions that ensured that the new DNA strand was radioactively labelled; this was done by making the new molecules from bases which had been made to contain a radioactive atom (see Box 5.1); such a radioactive DNA molecule is called a *gene probe*. When added to the filters it binds *only* to colonies carrying the targeted human beta-globin gene, so making just those containing the gene cells radioactive. This radioactivity was detected as spots on X-ray film; the position of the spot(s) on the film corresponded to the position of the colony on the original plate (Figure 5.5), so identifying the bacterial cells containing the

---

Box 5.1 The nature of radioactivity

The atoms of each of the elements known to chemistry co-exist in slightly different forms. Hydrogen, for example, has three sorts of atoms. The lightest and simplest in structure weighs one unit. A second version incorporates an additional sub-atomic particle, bringing its weight up to two units; in all respects other than weight it behaves exactly like the first version. The third possesses yet another particle and weighs three units. Chemically it is still every bit hydrogen, but this new configuration is unstable and tends to disintegrate into an atom of another element, helium, which also weights three units. The disintegration is accompanied by the emission at very high speed of a tiny, virtually weightless, particle called an *electron*, which carries a single negative electric charge.

In some respects this electron behaves like a ray of very energetic light; for example, it will fog a photographic film. It is because of these 'rays' that the term 'radioactive' is used for such disintegrating atoms. Every one of the elements (of which more than 115 are now known) has some radioactive varieties (called *radioisotopes*) among its atoms.

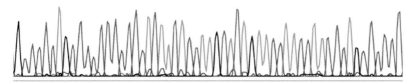

G T T C T C A G C T T C C T T C C T C A C A A C A T C A A G C A C A G A A T C A A T C A A C T C A G C T C C C T(

Figure 5.6 The ultimate objective of gene hunters is to elucidate the order of the bases in the DNA sequence of the gene of interest.

human gene of interest. The colony was grown up in a liquid broth to produce billions of cells and trillions of DNA molecules, all carrying the human gene for beta-globin.

In the good old days the scientist had to re-isolate the fragment by digesting the plasmid with EcoR1, purifying the inserted DNA and subjecting it to a stringent chemical analysis to work out the order of bases that constituted the gene. Nowadays one posts it to a laboratory specialising in sequencing. There they use high tech machines to work out the order of bases and produce a computer read out (Figure 5.6), which they e-mail back to the scientist who did the cloning!

The agony and the ecstasy
Considerable experimental skill and patience were required to clone the gene for the beta-globin protein. However, the fact that the defective protein had already been identified, and that lots of the appropriate RNA is found in red blood cells, facilitated the production of an appropriate probe. This was not the case with cystic fibrosis, Huntington's chorea, muscular dystrophy and a host of other debilitating inherited diseases. When the hunt for those genes began, all that was known about the genetic defects was that they existed. The nature of the protein involved in the diseases and the amount of RNA involved in making the offending protein was unknown, and in many cases scientists did not even know on which chromosome to search for the gene. Cloning genes without knowing the precise defect is incredibly difficult because there is no way of producing a probe that can identify the gene in the library of fragments.

The exact procedure used in these circumstances varied from gene to gene but much of it depended on using several restriction enzymes to detect changes in the DNA close to the gene being hunted. Sophisticated but painstakingly slow cloning procedures were deployed to search through millions of bases on either side of the detected change in the DNA to track down the genetic quarry. Each one of these projects required an enormous input of resources and frequently required international co-operation between large research groups. Each hunt took years, millions of pounds, dollars or whatever, and endless tenacity and patience. In every case, the agony of a whole series of disappointments gave way to ecstasy when genes were discovered thought by many to be beyond our abilities. Today scientists have succeeded in identifying and cloning the genes responsible for causing all of the major genetic defects arising from a *single* gene. As each becomes available, scientists and doctors are presented with their first real insight into the cause of the disease and the hope of eventually developing a treatment. But most importantly, the procedure provides information that can immediately be used easily to detect the defective gene in other individuals without having to resort to cloning their genes.

A family snapshot

Establishing the sequence of a cloned gene and its mutated counterpart facilitates the development of a test to discover which version of the gene is present in cells from any given individual. Although they vary in detail, such tests often exploit the fact that mutations are frequently associated with changes affecting the way the DNA within or close to the gene is cut by one or more restriction enzymes. For instance the base sequence CCTGAGG occurs three times in the gene encoding the *normal* beta-globin. The restriction enzyme MstII cuts at this sequence and therefore cuts this gene into two pieces (one is 200 bases long; the other is 1,100 bases long). In the mutant sequence, the A (underlined above) has been altered to a T in the central MstII site so the sequence reads CCTGTGG. Now, although MstII can still cut at the extreme ends of the gene, it cannot cut at the central sequence (the site of the mutation) because it no longer meets the enzyme's recognition requirements. The DNA from a person with sickle cell anaemia cut with this enzyme thus yields only one large fragment of 1,300 bases versus two fragments, one of 200 and the other of 1,100, characteristic of the normal gene. Despite the fact that there are hundreds of thousands of other fragments generated by a MstII digestion of total human DNA, recombinant DNA technology can easily pick up the difference in the fragments arising from the normal and mutant beta-globin genes.

The digested DNA is placed in a well at one end of a horizontal thin slab of jelly through which an electrical current is passed. The DNA molecules are carried through the gel by the electric current but their speed of movement is restricted by their size – large ones move more slowly than small ones. After a few hours, all the fragments are arranged along the gel *in*

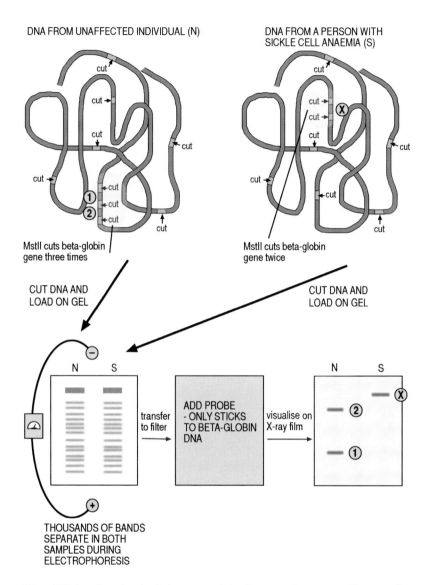

Figure 5.7 A radioactive 'probe' can reveal the fragments from a specific gene after digested DNA has been size sorted by electrophoresis.

the order of their sizes. The bands from the particular gene under study (beta-globin in this case) can then be visualised by transferring the DNA to a filter and adding a radioactive probe which binds only to the DNA of interest. The fragments are then detected on an X-ray film. In this case the probe reveals two bands of 200 and 1,100 bases for the normal gene, while the mutant yields give only one of 1,300 (Figure 5.7).

This technology can be used to follow the inheritance of genes and to detect the genes present in a foetus during pregnancy. Bearing in mind that we each possess *two* genes for each genetic trait, the hypothetical family depicted in Figure 5.8 illustrates how this works. Each parent has three bands indicating that they *both* carry one defective gene as well one normal one gene for beta-globin: the band of 1,300 bases comes from the mutant gene while the two at 200 and 1100 from the normal (Figure 5.8). Offspring 1 shows *only one band of 1,300 bases* – such individuals completely lack the normal gene and therefore have sickle cell anaemia. Offspring 2 has three bands indicating the presence of one normal and one mutant gene. Sample 3, from foetal cells, has only two bands which indicates that the foetus has two normal genes. This technology can be used on DNA samples from a whole series of individuals but it is expensive, makes high technical demands and can take days to complete. Thus, while it is an extremely useful *research* tool, it is of limited use for mass *screening* purposes. An alternative approach to cloning genetic information changed all this in the mid 1980s.

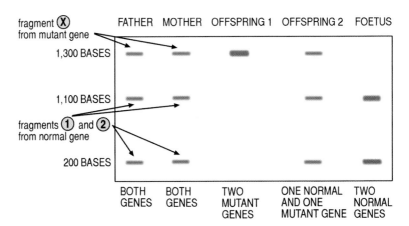

Figure 5.8 Detecting genes in a hypothetical family.

A revolution on the way to Mendocino

There are two types of developments in technology: the refinement of an existing technology and quantum leaps forward when the same problem is approached from an altogether different angle. Gene cloning was the culmination of a series of discoveries made in disparate fields and then brilliantly combined by the collaboration of two scientists who pooled their expertise to produce a breathtakingly exciting development in genetics. If this was an example of refining existing technologies, the invention of an alternative technology to reach the same objective was most certainly of the quantum leap variety.

Kary Mullis was preoccupied as he drove along Interstate 128. His mind was struggling to solve the problem of identifying the presence or absence of mutations in human genes. The year was 1983. As he juggled a number of disparate elements of research, an idea crystallised in his mind which was not the solution to the problem he had set out to address, but one that he realised immediately could revolutionise identifying invisible genes and much more besides. His idea was as brilliant as it was simple. He called it the *polymerase chain reaction* (PCR): it took him 3 years to perfect it in the laboratory, it earned him a Nobel Prize and the company he worked for a patent that they later sold for a third of a billion dollars! His procedure can produce millions of copies of any DNA sequence from any DNA sample (no matter how complex) in a matter of hours. It revolutionised gene visualisation and almost every other area of genetic engineering – its impact was so complete that, within a very few years, Mullis' work had become the second most quoted paper ever published in the field of biology!

When cells divide, they use an enzyme called *DNA polymerase* to copy their DNA template. Mullis eliminated the 3 billion base pairs of human DNA which were of no interest to his analysis by adding DNA polymerase to a test-tube of DNA and arranging for the replication (copying) of *only the gene he wished to examine*. By repeating this cycle ten times the target gene would be multiplied over 1,000-fold; continuing the reaction for a further ten cycles would increase the number of copies of the DNA of interest to more than a million. This serendipitous discovery of the PCR reaction did away at a stroke with the need for plasmids, restriction enzymes, the superglueing enzyme and even the humble bacteria.

The procedure is eminently useful for screening purposes. The only sample needed is a few cells obtained from the patients by a simple cheek swab. The cells are burst open to release the cellular DNA which is heated in order to separate the two strands. Short single strands of DNA (called 'primers'), which have been specially synthesised chemically to carry a section of the base sequence of either end of the gene in question, are added

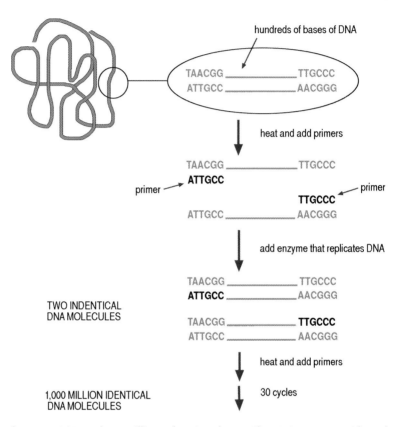

Figure 5.9 PCR produces millions of copies of a specific DNA sequence without the need for plasmids, restriction enzymes or bacteria!

to the cellular DNA where they bind to the targeted gene. When DNA polymerase is added, it recognises the short priming strands and replicates only the DNA lying between them because double-stranded DNA is required for replication and our DNA, having been heated, is double stranded only where the primers have bound (Figure 5.9).

A series of temperature changes is used uniquely to amplify the target and the entire process of gene amplification is completed within a few hours. A number of different analyses can now be performed on the DNA. In the case of sickle cell anaemia, it can be cut with MstII as before. There is no need to use radioactive probes to find the gene of interest because essentially the only DNA present is that of the target gene (it is the section amplified a

DNA FROM UNAFFECTED INDIVIDUAL

DNA FROM A PERSON WITH
SICKLE CELL ANAEMIA

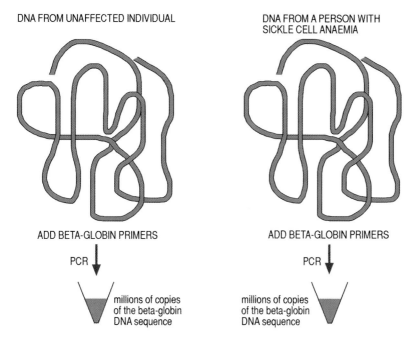

ADD BETA-GLOBIN PRIMERS

PCR

millions of copies
of the beta-globin
DNA sequence

ADD BETA-GLOBIN PRIMERS

PCR

millions of copies
of the beta-globin
DNA sequence

ALL FOUR SAMPLES CUT WITH MSTII

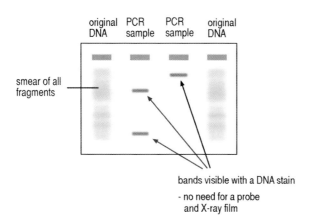

original
DNA

PCR
sample

PCR
sample

original
DNA

smear of all
fragments

bands visible with a DNA stain

- no need for a probe
  and X-ray film

Figure 5.10 PCR simplifies genetic screening because it is faster and does not require
radioactive 'probes'.

million times – the rest is not detected). Furthermore, there is now so much DNA present that a simple dye which attaches to DNA suffices to visualise the band patterns (Figure 5.10).

To use PCR in this way one must know the gene sequence of the target so there is still a definite need for the 'traditional' approach to gene manipulation. However, in a few short years the PCR reaction has revolutionised recombinant DNA technology beyond recognition because it is extremely sensitive, very fast, incredibly accurate and new applications of the basic technology have been continuously developed since its original invention. The technology has been commercialised to become simple to handle and first-year undergraduates use it in class experiments.

Dealing with the invisible is now so commonplace as to be unremarkable. Yet, even as I write this, I am acutely aware that the output from the most recent use of this technology is anything but. Read on!

6

Awesome analysis

They said it couldn't be done, that it was too costly even to try, that such information would be useless anyway. They were wrong. Less than 10 years on the task has been completed, the cost modest and the information generated invaluable. It has taken a century of research to get to this position. The effort has been tremendous; the potential fruits unimaginable. An awesome analysis, undreamed of less than 15 years ago, has recently revealed the entire genetic blueprint of a human being and, in doing so, provides an entirely unexpected answer to the ancient dictum 'Man – know thyself'.

The problem with humans

The vast majority of geneticists fought shy of human genetics for most of this century for very good reasons – humans were too complex, produced too few progeny, took too long to reproduce and they were not particularly amenable to controlled breeding programmes! Geneticists much preferred to work with other organisms (Figure 6.1). Some used mice which were considerably simpler, produced lots of progeny in a much shorter time span and raised no ethical objectives to the odd bout of incest. Others used the fruit fly because they were simpler still, demonstrated even greater fecundity than mice and had giant chromosomes, which were extremely amenable to microscopic analysis. A simple nematode worm was the choice of others. This animal consists of less than 1,000 cells (humans we noted earlier have 50 trillion), yet it does have simple, nervous, circulatory, excretory and reproductive systems, so providing a model for how multi-cellular organisms arise from a single fertilised egg. Still others concentrated on the genetic analysis of baker's yeast in an attempt to understand how single cells work. The principles used in the genetic analysis of these so called 'model organisms' are always the same: isolate individuals carrying inherited dysfunctions and try to work out the reason for the fault. This approach produced tens of thousands of mutated organisms, each one revealing some new aspect of how genes exert their influence.

While the geneticists working with organisms other than humans were

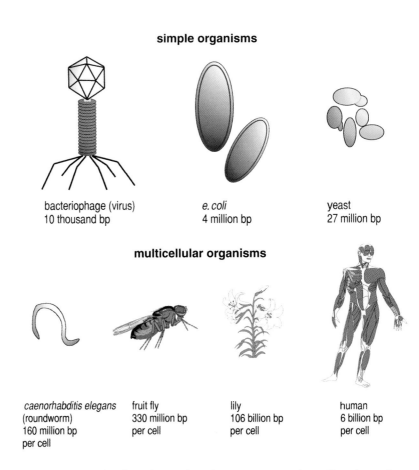

simple organisms

bacteriophage (virus)
10 thousand bp

e. coli
4 million bp

yeast
27 million bp

multicellular organisms

caenorhabditis elegans
(roundworm)
160 million bp
per cell

fruit fly
330 million bp
per cell

lily
106 billion bp
per cell

human
6 billion bp
per cell

Figure 6.1 Much has been learnt about how genes exert their effects by studying organisms that are less complex than man.

busy identifying thousands of proteins, cloning and sequencing the genes, encoding them and entering them onto databases, human geneticists were struggling with every gene they tried to identify. The research efforts needed to isolate, clone, sequence and analyse the genes responsible for even simple genetic defects such as sickle cell anaemia, cystic fibrosis, Duchenne muscular dystrophy and Huntington's chorea were nothing short of heroic. Years of effort and a good deal of money were needed to identify and characterise each one of these genes. Given the cost and effort required to isolate and analyse genes responsible for these easily recognisably inherited traits, how could geneticists hope to clone the other 75,000 genes, many of which are responsible for much more complex and widespread diseases like heart problems, cancer, diabetes and psychiatric disorders?

When all else fails . . .

A radically alternative approach to this problem was proposed in the late 1980s. Rather than having different groups attempting to isolate, clone and sequence individual genes, why not pool resources, 'blindly' clone and sequence *all* the genetic information in a human cell and deposit the sequence in a gigantic database for further analysis? It was argued that every one of the instructions needed to produce a human could be accessed from such a database and that it would become an invaluable research resource for decades to come.

Genome is the word used to denote all of the DNA sequences that an organism possesses. As creatures became more complex, they required more genes to encode the proteins for their structures, maintenance and reproduction. Multi-cellular organisms (the plants and animals we see all about us), not surprisingly, possess genomes substantially larger and significantly more complex than those of single-celled species like bacteria or yeast. The simple nematode worm, with a body containing fewer than 1,000 cells, requires 19,099 genes in order to function, whereas the assembly of a human body with 50 trillion cells needs an *additional* 55,000 genes which are not found in the simpler nematode. Complex genomes also contain large chunks of genetic material with no apparent role for encoding cellular proteins. This non-coding DNA, often called *junk*, may act as a reservoir from which new genes with novel functions can emerge, although nobody really knows for sure why it is there; it is so prevalent in higher organisms that only 90 million of the 3 billion bases in the human genome actually encode proteins! About 97% of our DNA is junk.

Exploring the entire genome sequence of any organism is an extremely demanding task because every bit of DNA in the cell has to be cloned, arranged into a logical order and sequenced. In the early 1990s DNA sequencing successes were measured in hundreds of thousands of base sequences *per year*; the idea of sequencing all *three thousand million* bases of DNA from a human cell seemed to many observers to be quite ridiculous. Other detractors complained that the complete DNA sequence was unnecessary; after all, 97% of it did not encode proteins. Furthermore, they argued, simply finding the DNA sequence for those proteins that were encoded would contribute very little information about the biology of the system. They maintained that the many billions needed even to attempt the project would be much better spent on more focused research programmes.

Figure 6.2 The human genome project is transferring the genetic blueprint for a human into a computer databank. Reproduced courtesy of DOE Office of Biological and Environmental Research, Human Genome Program USA.

Despite heavy resistance, a huge collaborative international effort aimed at producing the entire DNA sequence for a human being was initiated in 1990, with a projected completion date of 2005 (Figure 6.2). A number of impressive technological breakthroughs have speeded this up and the 'first edition' has already been released. For comparative purposes, the genomes of the model organisms (fruit flies, worms etc.), so beloved of geneticists working on non-human research programmes, were sequenced at the same time.

Reading the instructions
The instructions for setting up a video, or assembling euphemistically labelled 'build it yourself' furniture, may frequently be indecipherable, but they are positively explicit compared with the gobbledegook that appears on a computer screen when the DNA sequence for the assembly of a living organism is accessed. All that you see is an incredibly monotonous array of As, Ts, Gs and Cs in an apparently haphazard order – but of course the order is not haphazard. Somewhere in there are thousands of genes, each one with its own unique and highly ordered DNA sequence – it is simply a matter of finding them. Luckily every gene commences with the triplet sequence 'ATG'

and ends with one or more of the triplets 'TGA', 'TAA' or 'TAG' (the latter are the stop signals noted in Table 3.1). High-powered computer programmes use these and similar types of clues rapidly to deduce the presence of gene sequences amidst the endless millions of letters.

With the complete DNA sequence of an organism safe in a gigantic database, some or all of its genes can be compared with any other sequence (be it a single gene or an entire genome) already present in other databases. A yeast–worm comparison reveals that of the 6,340 genes present in the yeast genome, 3,000 are also present among the 19,099 genes in the genome of the worm. This essentially identifies the core of genetic information required to keep a cell alive regardless of whether it exists as a free-living individual cell or as any one of a large number comprising the body of a multi-cellular organism. When one recalls that a cell is a cell is a cell, there is every reason to predict that the vast majority of these seemingly core genes will also be found in the genome of every multi-cellular organism, including humans.

The 16,000 genes present in the worm, but not in yeast, are obviously necessary for dealing with the biological challenge of developing a single fertilised egg into a simple animal of about 1,000 individual cells living in harmony and co-operation with one another. Once again, bearing in mind that the different cell types found in this primitive animal are also found in humans, it can be confidently predicted that the majority of those 16,000 genes (or something very like them) will also be found somewhere in the 75,000-odd genes of the human genome (Figure 6.3).

It is obvious that a great deal of information can be obtained by simply reading and comparing genomes from different organisms. However, *reading* the information in an assembly booklet is one thing but, as thousands of unprogrammed video recorders and incorrectly assembled items of furniture testify only too clearly; *understanding* the instructions is something completely different again.

Understanding the instructions
At its most basic level, the genetic instructions of nuclear DNA are manifest as the proteins present within the cell. The precise function encoded in a gene is revealed only when the protein for which it codes has been identified. Luckily, the deep-rooted similarity of function in all living organisms means that particular gene sequences encode the same function no matter in which organism they occur – decode the gene function in one organism and you have almost certainly decoded it for them all. Yeast mutants have been generated and studied for decades, so the functions of many of their genes are already well known. This has allowed biologists who have never seen a

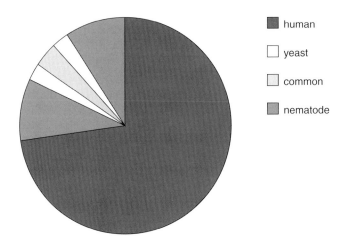

Figure 6.3 As organisms increase in complexity they retain genes that encode proteins with basic functions and accumulate genes encoding proteins with new functions.

nematode worm or extracted its DNA to use computer analysis to identify the functions encoded by hundreds of worm genes. They do so simply by calculating how closely a worm gene sequence resembles that of a yeast gene whose function has already been identified using traditional mutation studies.

The same is true for humans. Cancer arises because the normal control of cell division goes astray in some way, yet the basic model for the control of cell division was worked out by studying mutant yeast cells unable to divide properly. Not only are the overall principles the same but, in some cases, genetic information has been removed from human cells and inserted into yeast cells where it provides a sufficiently similar protein to replace the defective protein encoded by the mutant yeast gene. It looks as though, once nature has hit upon a good way of doing something, it keeps it in hand for future developments (Figure 6.3).

Such conservation of function has already seen the yeast genome sequence impacting on human genetics. *Rhizomalic chondrodysplasia punctata* is a rare, recessively inherited genetic condition which results in severe growth and mental retardation in those who inherit the mutation from both parents. The defective gene was isolated in 1997 and a computer used to compare its DNA sequence with all the genes of yeast. Surprisingly, a closely related gene sequence was identified, mutation in which had severe consequences for individual yeast cells. By using the normal human gene to

Table 6.1 The functions of many genes in human cells are closely related to those found in yeast cells. After Sudbery P. *Human Molecular Genetics*, Addison Wesley Longman Ltd, 1998.

Human disease (Symptoms)	Human gene	Yeast gene	Yeast cell defect
Cystic fibrosis (Defective membrane protein)	*CFTR*	*YCF1*	Defective membrane protein
Neurofibromatosis (Benign nerve cell tumours)	*NF1*	*IRA2*	Defect in a cell signalling protein
Ataxia telangiectasia (Poor co-ordination due to nerve degeneration)	*ATM*	*TEL1*	Defect in a cell signalling protein
Rhizomalic chondrodysplasia punctata (Severe growth and mental retardation)	*RCP*	*PEX1*	Defect in cell building protein
Hereditary non-polyposis colorectal cancer	*MSH2*	*MSH2*	Defect in a DNA repair protein

repair mutated yeast cells, this gene product was shown to perform identical roles in human and yeast cells. That is not an isolated example: Table 6.1 lists known human gene defects and their yeast counterparts. All of these relationships were found using database searches, obviating the need to use recombinant DNA technology to address these specific questions.

Of mice and men

One of the most outstanding attributes of humans is their high level of intelligence, so it is not surprising to hear that fully 50% of the 75,000 genes in the human genome are expected to encode proteins involved in the development, maintenance and regulation of the brain. Despite the fact that the yeast and nematode genome sequences have been and will continue to be extremely useful for identifying many human gene functions, they cannot shed much light on brain development and function. For this reason the consortium of world scientists decoding the human genome are also producing the entire genome sequence of the organism that has long become the mainstay of biological research – the laboratory mouse. The mouse genome is approximately the same size as the human and, although it is immediately obvious that some master and regulatory genes must differ between the two

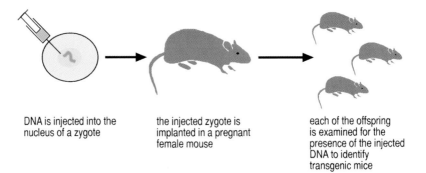

DNA is injected into the   the injected zygote is   each of the offspring
nucleus of a zygote   implanted in a pregnant   is examined for the
                                  female mouse   presence of the injected
                                                                  DNA to identify
                                                                  transgenic mice

Figure 6.4 Genetically engineered mice producing dysfunctional versions of proteins
that they share with man, can be used to understand human diseases and
to develop new drugs.

species, many human genes have already been identified in the mouse
genome using conventional recombinant DNA technology. Furthermore,
technology is now available allowing fertilised mouse eggs to be genetically
manipulated (Figure 6.4) to remove individual genes, thus facilitating the
analysis of their effects in mice. This so-called *gene knockout technology* has
already been used to produce mice suffering from all sorts of conditions from
which humans suffer, including diabetes, cancer, age-related blindness and
hardening of the arteries. Quite apart from their usefulness in the development of new drugs, the application of this technology to mouse brain development and function will provide deep insights into how the human brain
develops. To this end mice already exist that suffer from gene defects thought
to be involved in the development of Alzheimer's disease.

The human genome sequence will provide information to be used by
scientists for many decades to come as they attempt to solve the mysteries of
how genes affect us in health and in disease. But it will be an equally valuable tool in the area of genetic diagnostics.

Personalised analysis
The human genome project commenced at a number of centres throughout
the world with each group opting to sequence a different chromosome.
People used the most readily accessible source of DNA; despite the fact that
the project is referred to as *the* human genome project, the first human
genome sequence will in fact be a compilation of genetic information from a
number of different individuals. But, as the average difference between any
two human genomes is no more 1% of the 3% which is not junk (i.e. only

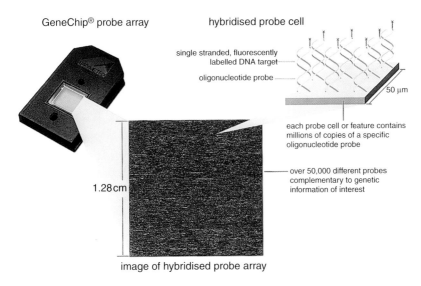

GeneChip® probe array

hybridised probe cell

single stranded, fluorescently labelled DNA target

oligonucleotide probe

50 µm

each probe cell or feature contains millions of copies of a specific oligonucleotide probe

over 50,000 different probes complementary to genetic information of interest

1.28 cm

image of hybridised probe array

Figure 6.5 They might only be a few square centimetres in size but GeneChips® are revolutionising how we analyse genes. Reproduced courtesy of Affymetrix Inc.

0.03% of the total, junk included), *the* human genome is sufficient for most basic research purposes. As the sequence comes on stream, scientists will be able to concentrate on those regions of the DNA which actually encode proteins. At the leading edge of these investigations will be a new technology which is already changing the way we examine gene sequences and promises a future of ever increasing accuracy and speed in the detection of defective genes. GeneChip® probe arrays (Figure 6.5) are only a few square centimetres in size but their power of analysis is remarkable.

In Chapter 3 we saw how DNA is a molecule consisting of two complementary strands. Referred to as the *sense* and *anti-sense strands*, the As in one strand are always joined to the Ts in the other, while Gs are always joined to Cs. Mutant genes have a slightly different base sequence from that of the normal version of the gene. Therefore, if the sense strand from a mutant is mixed with the anti-sense strand from the normal gene there will be a discontinuity in the helix (Figure 6.6) which, under certain conditions, can be accentuated *to prevent any binding at all.*

This is the basis for detecting mutations using chip technology. Minuscule amounts of tens of thousands of short DNA strands, each comprising 20 bases of the sense strand from a known gene or genes, are arranged by a

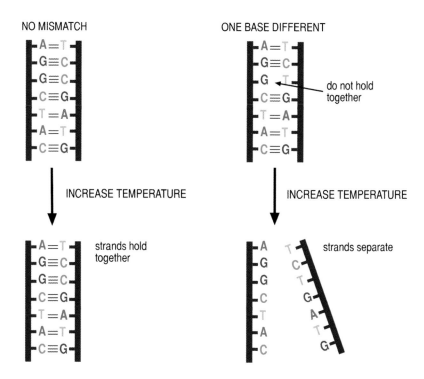

Figure 6.6 A single base change can prevent two otherwise complementary DNA strands from annealing to one another.

robot on a single microscope slide. The DNA from the individual to be tested is isolated, fragmented into sections of about 100 bases long and stained with a green fluorescent dye. DNA from an individual known to possess the *normal* genes is also isolated and fragmented but stained with a red fluorescent dye. Both samples are then placed onto the GeneChip® probe array. If the patient has a mutation, their DNA will bind less effectively than the 'normal' DNA. A laser detector scans the chip and, after checking each position on the grid, signals whether it is green or red to a computer. A computer read-out of fluorescence patterns reveals the position of any mutations in the patient's DNA. This offers for the first time a screening procedure that can simultaneously detect all possible mutations in a particular gene.

Several GeneChip® probe arrays are already available for research purposes, including one to detect mutations in the gene giving rise to cystic fibrosis, another to spot mutations in the beta-globin gene and yet others to find mutations in various genes known to be associated with cancer. This so-called 'microarray technology' is just starting to come on stream and is already perceived to be as accurate and sensitive as more traditional methods, yet requiring less laboratory effort and offering greater flexibility and sensitivity. The present GeneChip® probe arrays carry up to a million different sequences; this figure is increasing continually, with the confident expectation that, over the next few years, sequences capable of detecting mutations throughout the entire protein-coding region of the human genome will fit onto one of these chips. It took 12 years and many millions of pounds to sequence the first human genome. In the near future it should be possible to obtain the sequence of the protein-coding regions in our own genomes at a negligible cost within hours of supplying a few cheek cells to the laboratory.

Unearthing the information in individual genomes will have enormous practical value, particularly for healthcare – but will it reveal so much about every man, woman and child that their futures will amount to little more than the information contained in their genes at the moment of conception? Will individual genome sequence information provide a complete answer to the ancient dictum with which we started this chapter, or is there more to an individual than the genetic information that determines the structure and function of his or her body?

Much more than a genome

Phenylketonuria is a recessively inherited disease causing severe mental retardation. Until relatively recently, individuals who inherited a copy of this gene mutation from both parents were doomed to spend their days in some sort of institution. Today, people carrying the very same mutation can lead a perfectly normal life. The affected individuals have a mutation in a gene encoding an enzyme used for the metabolism of phenylalanine, a naturally occurring amino-acid and a normal constituent of our food. In the absence of this enzyme, the amino-acid accumulates in the bloodstream and another enzyme, in a forlorn attempt to remove it from the body, inadvertently converts it into a molecule which is particularly poisonous to nerve cells in a developing brain. The ensuing brain damage is the source of the mental retardation.

All babies are now screened at birth for this enzyme deficiency by a simple blood test. Individuals lacking the enzyme are kept on a strict phenylalanine-free diet for 7 years until their brains have developed; as a result, this type of mental retardation has now disappeared. The gene mutation is still present in the population, the enzyme deficiency is also apparent in the infants affected but, as long as affected individuals avoid phenylalanine, their nerve cells are safe and the gene mutation has no noticeable effect. What such an individual becomes in life is not determined just by the gene

Table 6.2 The manifestation of the recessively inherited condition phenylketonuria in humans depends on the diet of the affected individual. After Sudbery P. *Human Molecular Genetics*, Addison Wesley Longman Ltd, 1998.

	Medical condition	
Genetic constitution	Average diet	Phenyl alanine-free diet
2 NORMAL GENES	Normal	Normal
1 NORMAL and 1 MUTANT	Normal	Normal
2 MUTANT GENES	Mentally retarded	Normal

combination in his genome – the local environment of the individual cells in that body is at least as important (Table 6.2)!

If the gene sequences in an individual's genome were the main determining factors in a person's life, one would expect identical twins (with identical genomes) to lead nearly identical lives and suffer from identical diseases. Identical twins can certainly look very similar and act very much alike, but extensive research on identical, as opposed to non-identical (i.e. those with different genome sequences) twins, reveals that they lead anything but identical lives. Table 6.3 shows that although there does appear to be a greater tendency for identical, as opposed to non-identical twins, to suffer from the same diseases, there is also a large non-genetically determined aspect to all these conditions. Psychiatric disorders are more likely to develop in the second identical twin if the first develops the condition, but it is by no means certain. Even performance in intelligence tests, although heavily affected by an individual's genome sequence, is in sizeable proportion determined by non-genetic factors. However, even in such cases, twin studies highlight the fact that, whereas a particular combination of genes might *predispose* an individual to a particular condition, it certainly does not determine that it will definitely occur. This is perhaps most forcibly brought home by the surprising observation that 93% of breast cancer cases arise without any obvious genetic predisposition to the disease. It appears that the interaction of genes and their environment are never quite the same in any two individuals; this is not really surprising when one considers the intricate and incredibly complex network of interactions that give rise to each one of us.

Foetal development is an extremely dynamic and complicated process. All sorts of physical factors impact on a baby in the womb and even a slight difference in blood supply is reflected in a different rate of foetal growth. Given that there are five billion nerve cells in the developing brain, it is not difficult to see how mental abilities might vary between identical twins even from the moment of their birth. Differences can only increase as the babies

Table 6.3 Comparison of identical and non-identical twins reveals a significant level of non-genetically determined influences on a wide range of human conditions. After Sudbery P. *Human Molecular Genetics*, Addison Wesley Longman Ltd, 1998.

Trait	Number of pairs exhibiting the trait	
	Identical twins	Non-identical twins
Late onset diabetes	100	10
Breast cancer	7	6
Insulin dependent diabetes	50	5
Tuberculosis	51	22
Alcoholism	40	20
Alzheimer's disease	58	26
Dyslexia (reading disability)	64	40
General intelligence	80	56
Extrovert behaviour	50	18

get older because they will encounter and learn from varied life experiences as they start to explore the world around them.

Second in complexity only to the brain is the immune system. This also develops throughout the life of an individual as it responds to the microbial invaders encountered in the environment. Again, subtle differences arise in the immune systems of identical twins depending on the environments which they encounter. As many chronic diseases, including insulin-dependent diabetes, rheumatoid arthritis and allergies, result from overactive immune systems it is easy to see why identical twins might suffer from different types of diseases. Chapter 4 pointed out that lung cancer has little to do with genes and everything to do with smoking; there is a similar story with the type and quality of food we ingest and our tendency to develop heart disease. It therefore goes without saying that there is much more to individuals than the genes they are born with!

Perhaps it is not surprising that the assembly of 50 trillion cells from a single fertilised egg should have a certain flexibility built into it, as a direct consequence of which no two individuals are ever identical – even when they do arise from cells containing the same set of genetic instructions. It is somehow reassuring to realise that nature plays the role of an uninhibited artist rather than that of a rigidly disciplined technologist as each individual completes his or her incredible journey from a fertilised egg containing a single genome of information to a multi-billion celled organism capable of decoding it in its entirety.

7

Knowledge is power

Detailed knowledge about the genetic constitution of an individual provides information totally unlike anything accessible from any other source. Not only can it be used uniquely to identify people and inform them about the genes they currently carry, but it can also anticipate the likelihood of particular genes arising in their offspring. Furthermore, it connects an individual with those ancestors from whom the information has been inherited, thus providing a window onto the past. The acquisition of personal genetic knowledge empowers an individual but in so doing can have ramifications for parents, relatives, offspring and society at large. The power of these uniquely powerful data will require enlightened legislation to ensure that, from healthcare to forensic science, they are used appropriately.

A double-edged sword

Huntington's chorea is a devastating neurological disorder causing premature senility, dementia and eventual death of the patient. The symptoms of this dreadful disease manifest only in people in their late middle age, after they have had their families. It is caused by a dominant mutation and thus *one* copy of the gene is sufficient to cause onset of the disease; such an individual will have one normal gene and one mutated gene. Thus, each one of their offspring has a 50% chance of inheriting the defective gene and with it the certainty of developing the disease as they become older.

Consider then the case of a woman aged 25 whose paternal *great-grandfather* had the disease but whose paternal *grandfather* died at the age of 30 before he could reveal whether or not he had inherited the condition. Her father is in his forties and normal to date. However, because the disease does not manifest until late middle age, he does not yet know whether or not he inherited the mutated gene from *his father*. If he has, then there is a 50% chance that his daughter also has the gene. Aware of the devastating effect the disease had on her great-grandfather and knowing that there is no treatment available, would the 25-year old woman opt for a DNA test to see if she has been unfortunate enough to have inherited the gene? In the vast majority of real cases the answer is 'no'. Faced with an incurable/ non-preventable condition many, perhaps most, people would simply rather not know.

Now let us suppose that the 25-year old meets a young man in his mid-thirties and wishes to settle down with him. Knowing that he wants to start a family as soon as possible, she secretly gets screened for the Huntington's mutation. If she does not have it, all is well; if the test is positive she faces a number of extremely difficult decisions. Should she warn her future husband:

- that she is destined to develop a devastating disease shortly after their twenty-fifth wedding anniversary?
- that if they decide to have children, half of them will also inherit this dreadful disease?

But ironically this is the easier part of her dilemma. Diagnosis of such a mutation is completely unlike the type of medical diagnosis that doctors make on the basis of non-genetic tests. The latter apply only to the patient; the former will have implications for several (even many) people, some of whom may not wish to know how likely they are to develop a disease. The only person from whom she could have inherited the disease is her father, but he has never been tested and does not want to know if he has the defective gene. Should she tell him that he will definitely develop the disease? If not, should she tell his brothers and sisters (her uncles and aunts), each one of whom has now been shown to have that 50% chance of possessing the gene and going on to develop the disease. Some of those aunts and uncles might already have offspring; should she tell them, too? What about her own brothers and sisters? Each one of them also has that 50% chance of having the defective gene. Now we can begin to see just how powerful genetic information can be and the sort of ethical dilemmas it can pose even within a family.

The technology which allows the detection of the Huntington's chorea gene can free individuals and their entire families from the fear of developing a dreadful disease if it is negative. But if postive, the results from the same test can condemn the individual to years of anxiety and the other members of the family to years of doubt. Genetic tests do indeed provide individuals with choices but their attitudes to such tests have ramifications far beyond the person wishing to undergo them.

Agonising choices

Knowing about dominantly inherited conditions affects present and future generations as in the case of Huntington's chorea; information about recessively inherited diseases almost always impacts only members of the next generation. We saw in Chapter 6 how, unlike genetically *dominant* diseases, individuals develop *recessively* inherited diseases only after they inherit copies of the mutated gene from *both* parents. We all carry a number of recessive mutations but are unaware of them. Furthermore, the chances of having a family with a partner carrying the same recessive gene mutation is remote in the vast majority of cases. The fact that two individuals have the same mutation normally comes to light only when they have an affected baby. Frequently genetic tests are used to assess the genetic status of future babies while they are still foetuses in the womb. In these cases, the chance of the baby inheriting the defective gene from one parent is 1/2. This is serious only if the baby also inherits the defective gene from the other parent (again a chance of 1/2); the chance that the baby will inherit both defective genes is therefore $1/2 \times 1/2 = 1/4$.

When parents are at risk of having a baby who will suffer from a genetic disease, a minimally invasive technique is now available 10 to 12 weeks into the pregnancy to remove a tiny piece of the tissue covering the developing foetus. The cells comprising this membrane are foetal in origin and can be used to screen the foetus for a number of possible gene defects (Figure 7.1). When the tests show that the foetus does not carry the defective gene in question, the parents are relieved of 6 months' of anxiety. But if the baby will be born with a serious genetic defect, the prospective parents are faced with the agonising choice of seeking a therapeutic abortion or of going ahead with the pregnancy and mentally preparing for the birth of the baby. There are no easy answers to this highly emotive issue. The seriousness of the condition, the possible availability of treatment to provide the sufferer with a good quality of life – not to mention the ethical and religious background of the parents – all play major roles in making this decision.

Less contentious than therapeutic abortion is the use of polymerase chain reaction (PCR) technology for pre-implantation screening. When a couple have had a baby suffering from a serious inherited mutation, they can avoid having further affected individuals by *in vitro* fertilisation. Here the woman is given hormone therapy to induce the production of numerous eggs. These are fertilised with her partner's sperm *in vitro* (that is, outside her body) and the fertilised eggs start to divide in the laboratory. After three cell divisions there will be eight cells (Figure 7.2). One of them can be removed and the DNA it contains analysed for the mutant gene using PCR (see Chapter 5). If the normal gene is present, the small ball of the remaining seven cells is implanted in the mother and the foetus then develops quite normally. Whenever the mutant gene is found, the cells are discarded. While this type of technology is still ethically contentious, it avoids the more extreme measure of abortion towards the end of the first trimester.

1. ASPIRATE A SMALL SEGMENT OF TISSUE

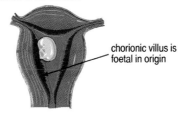

chorionic villus is foetal in origin

2. PCR UP TARGET GENE

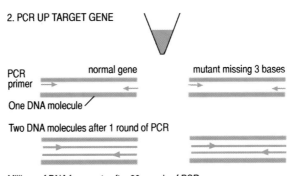

PCR primer

normal gene

mutant missing 3 bases

One DNA molecule

Two DNA molecules after 1 round of PCR

Millions of DNA fragments after 30 rounds of PCR

3. MAKE DIAGNOSIS
e.g. the cystic fibrosis gene versus normal gene

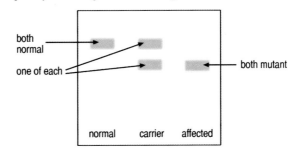

both normal

one of each

both mutant

normal   carrier   affected

Figure 7.1 PCR analysis of a tiny piece of tissue can reveal the genetic constitution of a developing foetus.

1. *IN VITRO* FERTILISATION

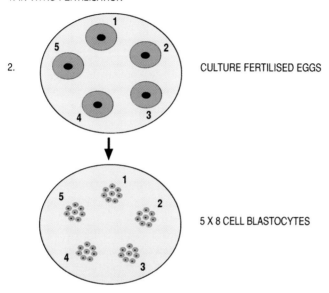

2.   CULTURE FERTILISED EGGS

5 X 8 CELL BLASTOCYTES

3. REMOVE ONE CELL FROM EACH

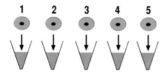

4. PCR AMPLIFY ALL THE DNA

5. THEN USE PRIMERS TO SELECT <u>ONE</u> GENE

6. USE A GEL TO IDENTIFY WHICH BLASTOCYST TO IMPLANT

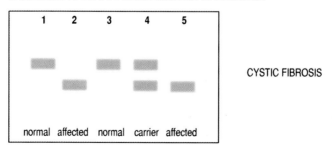

CYSTIC FIBROSIS

Figure 7.2 A combination of *in vitro* fertilisation and PCR can be used to avoid having a baby suffering from a genetic disease.

A foolproof method for avoiding the birth of babies carrying an inherited recessive trait is for prospective partners to submit themselves to DNA analysis before starting a family. Couples sharing a common mutation could avoid having children. Despite the inescapable logic of this argument, genetic counselling in practice has little or no effect on the mating patterns of individuals! However, with the notable exceptions of the genes responsible for sickle cell anaemia and cystic fibrosis, a screening programme would be of limited use because the chances of two individuals carrying the same mutation is small in a *randomly mating* population. Nevertheless, descendants of an individual carrying a particular trait have a much higher chance of carrying that trait so that genetic abnormalities often occur in so-called *inbred* populations. For instance, each of us carries 4–5 deleterious recessive genes. Suppose 1 person in every 100 carries a particular recessive; with random mating, the chance of two carriers having a child is $1/100 \times 1/100 = 1/10,000$ and, as there is only a 1 in 4 chance of having an affected baby even if both parents are carriers, the overall frequency of affected individuals is 1/40,000. If, however, marriages take place between first cousins, the chance of having an affected individual is 600 times higher at 1 in 64. Marriages within a particular religious or ethnic group tend to pose a similar (albeit lower) risk of genetic defects. Genetic screening programmes could be of benefit here because the types of inherited defects are known.

An effective genetic screen
One particularly informative example of the use of a genetic screen *before the event* is that for Tay Sachs disease among Ashkenazi Jews. Tay Sachs is a disease characterised by very early onset, progressive retardation in development, paralysis, dementia and death by the time the child is 3 or 4 years old; it is quite awful and to be avoided if at all possible. It is found in a French-Canadian isolate in Quebec, but is best known for a traditionally high incidence (1 in 3,600 infants) among Eastern European Jews amongst whom the carrier frequency is 1 in 20 to 1 in 30 (it is thought that the mutation occurred in Poland some 500 years ago). Among the Ashkenazim, some religiously ultra-orthodox communities practice arranged marriages, so all their children are scanned for Tay Sachs well before they might marry and the results are kept in a confidential register. When the matchmaker gets to work on a prospect, she (matchmakers are usually women) consults the register: if there is any risk of Tay Sachs, that partnership will not be pursued.

The health of the individual
So far we have considered the use of genetic screening procedures only to

identify individuals carrying recessive genes for serious genetic diseases. Such procedures are of limited use, however, not only because they can be used only to plan the *next generation* but also because the number of diseases caused by single gene mutations is quite small. It would be much more helpful for genetic screening programmes to identify individuals in the population who were prone to particular types of more common diseases.

We have already seen that there is a strong genetic component to many of the commonest diseases, including heart conditions, diabetes, mental disorders and cancer. We have also seen that their complex inheritance patterns arise because they are under the influence of several genes and there is also a large environmental input. Nevertheless, the web of genetic interactions conveying a predisposition to contracting the diseases is rapidly being identified (Table 7.1) and screening programmes will soon be in place capable of identifying individuals with a genetic predisposition to various types of illness. In fact, sequencing technology is developing so rapidly that, in the not-too-distant future, each individual will be able to possess their total genetic profile.

By identifying individuals predisposed to developing a disease early in life, preventative measures can be taken to improve lifestyle, ensure regular clinical monitoring and provide early medical or surgical intervention. Screening programmes will provide details about particular gene mutations but, given the complex origins of these diseases, the predictive value of the tests will not be absolute because environmental factors play such an important part in their aetiology. Here we catch a glimpse of a real problem: would you *really* want to know at an early age that there was a 30% chance of your getting cancer in your thirties or forties and that there were a number of things that you could do to minimise but not eliminate such a possibility? Would such information not hover like a dark cloud at the back of your mind when you were making career choices, marriage plans, starting a family? If

Table 7.1 The number of human genes that are being identified as contributing to common diseases is continually increasing. After Strachan T. *The Human Genome,* Bios Scientific Publishers Ltd, 1992.

Predisposition to	Gene(s) located on
Lung cancer	Chromosomes 1, 3, 5, 8, 17
Colon cancer	Chromosome 5, 17, 18
Insulin dependent diabetes	Chromosome 6
Coronary heart disease	Chromosomes 4, 12, 19
Alzheimer's disease	Chromosomes 19, 21

there were positive and helpful measures, definite steps to be taken seriously to diminish the risk, many people might opt for the knowledge, but suppose there were nothing to be done? At one level this type of information is of a totally personal nature but at another it could have far-reaching socio-economic consequences.

The results of genetic tests are unique in that they have as much to do with a person's medical *future* as they have to do with his or her medical *history*. As we accumulate more knowledge about how our genes affect our health and as micro-array technology becomes commonplace, it will become possible to access genetic information which an individual might wish to keep private: information which, if known, could affect the attitude of a potential employer or insurer towards that individual. Some argue that the results of such tests should remain confidential, others (insurers amongst them) maintain that the results should indeed be available because unscrupulous individuals could use this information for their own ends, perhaps to defraud insurance companies.

A fear that the results might somehow get into the 'wrong' hands stops many people from submitting themselves for testing, so they fail to find out about conditions possibly vital to their own welfare. This in turn undermines the extremely useful development of preventative medicine, with all the potential consequences on already over-stretched health service providers. The same fear also stops individuals from participating in drug trials in which researchers are looking for a genetic basis for the side effects certain drugs have on subsections of the population. Lack of subjects in turn delays finding the most effective drug treatments.

Enlightened and imaginative legislation is now required to address the potential uses to which an individual's genetic profile can be put. Given human nature and the increasing access to information seen in so many areas, it is probably best to minimise the amount of completely private information and ensure instead that legislation prevents discrimination against individuals with a genetic predisposition to various illnesses. Openness, coupled with anti-genetic discrimination legislation, might encourage the widest possible use of screening opportunities while at the same time assuaging the anxieties of insurance companies. We all share a common genetic heritage and, in many ways, we are all jointly responsible for the natural environment that impacts so many health issues. Every individual harbours potentially disadvantageous genetic information but only the unlucky few ever find this out. A flexible legislative programme leading to a uniform underwriting of the population by *all* members of society, including those deemed to be genetically healthy, has a lot be said for it.

Proof positive

While knowledge about our genes empowers us in so many ways, the information contained in that 97% or so of our DNA which does *not* code for proteins is equally valuable in providing a kind of understanding with a very different impact.

PCR can be used to amplify specific regions in our DNA which do not encode protein but simply consist of repeated bases, e.g. CACACA. When the measurements are done, the number of CAs on a chromosome a person inherited from his or her father is frequently found to be different from the number of CAs at the corresponding site on the chromosome from his or her mother. For instance, a site on chromosome no. 1 inherited from father might have two repeats (CACA), while the one from mother might have four (CACACACA). Children of such parents will inherit *either* a chromosome carrying two or a chromosome carrying four of those CA repeats. The paternity status of a child can be thus explored by measuring the number of CA repeats he or she has, compared with that in his or her father's DNA at that particular site. Note that the analysis is equally valid for determining maternity but usually that is in less doubt!

There are many of these variable sites scattered on different chromosomes throughout the genome, each with 10–20 possible repeat lengths and each one uniquely identifiable by PCR (four pairs of sites with 10 options at each gives 100 million different possible combinations). This makes it is feasible to build up a unique profile of bands for any given individual. Thus, a person's band pattern is unique and intimately related to his parents, offspring and all other members of his extended family. This technique is incredibly powerful for tracing relationships and has been used both to determine paternity cases and resolve immigration disputes.

Genetic anthropology

In Chapter 3 we saw how the sex chromosomes (pair no. 23) are special, not a pair in the same sense as the other 22. In females carrying two X chromosomes swapping is possible but in males with one Y and one X it is not. Because the Y chromosome cannot readily exchange any of its substance with X, the Y of each living man directly resembles that of his father, his grandfather, great-grandfather and so on back into history, undiluted by any interchange with chromosomes from the maternal line. Analysis of Y chromosomes is a very powerful way of looking at historical male lineages and the relationships between contemporary living males. Nevertheless, Y chromosome analysis, enormously valuable and significant though it is, reveals only a small part of an individual male's genetic make-up because Y is a very small chromosome and contains no more than about 2% of a man's total DNA.

There is also a way of looking at maternal inheritance. In addition to those 23 pairs of chromosomes, there is another place in cells where DNA is found: in tiny bodies called *mitochondria*. Although they are present in both eggs and sperm, the mitochondria remain outside during the act of fertilisation when the sperm's chromosomes are inserted into the egg. So all the offspring's mitochondrial DNA comes from mother and none from father.

The transmission of Y chromosome and mitochondrial DNA from generation to generation is shown in Figure 7.3: the blue lines show Y chromo-

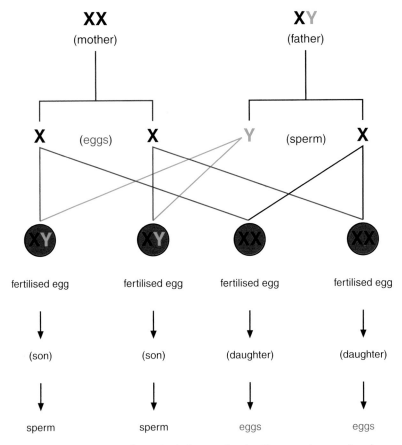

Figure 7.3 Some genetic information belongs uniquely either to males or to females.

some DNA passing from father to son, but never from father to daughter. The red lines from the mother show that only maternal mitochondrial DNA is passed to the next generation and that this gets transmitted to both sons and daughters (in addition to the X chromosome). However, as mitochondrial DNA is inherited via the egg, only the daughters transmit it to the next generation.

Easter Island

Mitochondrial and Y chromosome DNA analyses have thrown very interesting light on history and archaeology. Do you remember Thor Heyerdahl and his raft *Kon Tiki*? Heyerdahl wanted to test the notion that the inhabitants of Easter Island, a remote and isolated spot in the eastern Pacific Ocean, thousands of miles from the nearest land, had come originally from South America. They would have had only rafts for ocean travel so *Kon Tiki* was built to prove that it was possible to sail with such a vessel from South America to Easter Island. Heyerdahl succeeded– and for many years it was commonly held that that was the route taken to Easter Island. Alas for genetics! Recent analyses of mitochondrial DNA from people living in Melanesia, Micronesia, Easter Island and Southeast Asia have shown conclusively that the Easter Islanders migrated east across the Pacific, originally from the Asiatic mainland, just like all the other island populations.

The Jewish priests

Other studies of Y chromosome analyses follow paternal lineages; patterns of genetic information (see page 99) in the non-coding region can be explored in the search for ancestral relationships. There is a Jewish tradition that the sons and following male descendants of Aaron (the brother of Moses and the first High Priest) became members of a priestly clan called *Kohanim*; many people alive today belong to this group and some of them carry the name Cohen (or a derivative of it). Sons of Kohanim were told by their fathers that they were such members and they in turn told their sons. You could not join: the only way to be a member of the group was to be the son of a member. Whether or not the mother is the daughter of a Cohen is not relevant for her sons' affiliation.

Over the centuries since Aaron, many people, men included, have converted to Judaism but none of them could become Kohanim – joining is not allowed! So if the oral tradition had indeed been followed, one would by now expect to find evidence of a different genetic pattern distinguishing male Kohanim (all originating from Aaron) from other male Jews (originating from multiple sources): only true Kohanim would have inherited that Y chromosome pattern going all the way back to Aaron. Once more genetics proved the point: samples taken from several hundred Jewish males showed that there really is a difference and that the tradition does have substance. Further work sought to date the origin of this particular Y chromosome ancestry; it turned out to have been 2,600 years ago give or take a bit, just the right time between the Exodus from Egypt (during the reign of Ramesses II [BCE 1304–1237]) and the destruction of the First Temple (BCE 586).

Exploring our ancient genetic history

Genetic profiling technology also impacts on samples from the past. DNA is a particularly tough molecule. It does not break down easily and under certain conditions can be preserved for many thousands of years, so that DNA can still yield a genetic profile long after a body has decomposed. (*Jurassic Park* notwithstanding, the oldest sample so far yielding data from DNA is that from a Siberian mammoth frozen in the permafrost for somewhere between 50,000 and 250,000 years). Thus, in addition to assisting in establishing the identity of unknown corpses, DNA has participated in undertaking historical investigations. A dispute over the authenticity of a skeleton claimed by some to be that of the Nazi war criminal Josef Mengele was finally settled by comparing the genetic profiles of Mengele's wife and son with that of a small tissue sample from his purported skeleton. After eliminating the bands common to both son and mother, the remaining bands clearly matched the son to the skeleton so proving unequivocally that the skeleton was indeed authentic. Likewise, PCR-based genetic profiling on minute pieces of tissue from skeletons and on blood samples from existing relatives provided conclusive evidence that the skeletons recovered in Yekaterinburg were those of the Romanov family assassinated in 1918, allowing the remains to be accorded an appropriate burial in St. Petersburg together with other members of the Russian royalty.

One of the reasons that profiling technology is so powerful at identifying human remains is that PCR specialises in amplifying DNA. Extremely tiny quantities of fragmented DNA are sufficient for PCR to provide accurate information on the identity of an individual. It is for this reason that genetic profiling is invaluable and often at its most impressive when providing forensic evidence in criminal investigations. Such cases often offer no more than a drop of blood, a human hair, a swab or a semen stain; even a fingerprint might suffice. In the past very little firm evidence could be acquired from material like this but PCR has changed all of that. Now such samples play a leading role in protecting innocent individuals and in finding the perpetrators in rape and murder cases.

Nevertheless, and despite the impressive power of identification possible through genetic profiling, it has over the past 10 years been the subject of occasional fierce courtroom debate. Arguments about the accuracy of the experimental procedures and the variation seen in genetic markers between different ethnic communities have led to many instances where the profiling evidence has been thrown out of court. On even a quite basic level, O.J. Simpson's lawyers had the profiling evidence dismissed as inadmissible in his widely publicised trial because of inconsistencies in the way the forensic evidence had been collected and processed *prior* to its arrival in the laboratory.

However, each setback has simply resulted in the development of more stringent procedures and tighter quality control so that the technology is now extremely robust and reliable, a degree of reliability which will increase even more in the near future when micro-array technology meets the human genome project. Until then the quality provided by PCR technology is still extraordinarily high as William Jefferson Clinton, the most powerful man in the world, found out to his cost at his trial!

Recombinant DNA technology is remarkably efficient at determining both an individual's genetic identity and any genetic defects they might have. Identifying defects, however, is one thing – rectifying them is a totally different matter. That requires DNA to be manipulated so that its *information content* becomes expressed as a *functional protein* in the right place at the right time. Two quite different approaches are currently being pursued to use recombinant DNA technology for the alleviation of disease. One provides proteins of medical importance; the other attempts the wonder of repairing the genes themselves.

8

Marvellous molecules

The information content of human DNA is accessible to scientists because the simplicity of DNA made gene manipulation possible and facilitated the development of automated gene sequencing technology. The chemicals assembled into DNA are the same in all organisms and that is why the structure and function of DNA is the same regardless of the complexity of the organism in which it occurs. Variations exist between living organisms not because of differences in the *chemical structures* of their genes, but because the *order of the bases* in the DNA gives rise to unique combinations of proteins in each organism. Cutting and joining DNA is one thing; working out how to persuade a cell to use the information from a recombinant DNA molecule to make a protein is a different proposition. Having mastered basic gene cloning procedures, scientists next had to rise to the much more demanding challenge of how to put a human gene into a non-human cell and have this foreign genetic information decoded into a properly folded, functional protein.

Why bother?
Genetic engineering for the production of human proteins had a practical value beyond fundamental research. The well-being of a multi-cellular organism such as man depends on the proper functioning of its many tens of thousands of proteins. The majority of them are inside cells or stuck in the cell membranes. The rest are outside the cells in the fluid surrounding them and, of course, in the blood that supplies them. We have seen that many disease states arise because of defective proteins, but a brief glance at therapies shows them to be aimed exclusively at proteins in the bloodstream – in a sense, treating the symptoms, not the underlying cause. Diabetics receive insulin injections, haemophiliacs get blood clotting factors, growth hormone is administered for growth retardation, anaemia is treated with erythropoietin (EPO – see below) and so on. These proteins were available before the invention of gene manipulation but their sources were less than ideal. Insulin was isolated from pigs' pancreas; clotting factors came from litres of blood donors' plasma; growth hormone was extracted from the pituitary gland at the base of the brain of human cadavers, while urine was

the source of EPO. These preparations had their shortcomings: blood and cadaver products carry with them the distinct possibility of viral infections, while animal proteins pose the additional risk of being recognised as foreign by the immune system and inducing a severe immune response in the recipient.

Genetic engineering was seen as the ideal solution to all of these problems. If human genes encoding these proteins could be cloned into plasmids in *E. coli*, the bacterial decoding machinery could be used to provide a safe, relatively inexpensive and limitless supply of precious proteins (Figure 8.1).

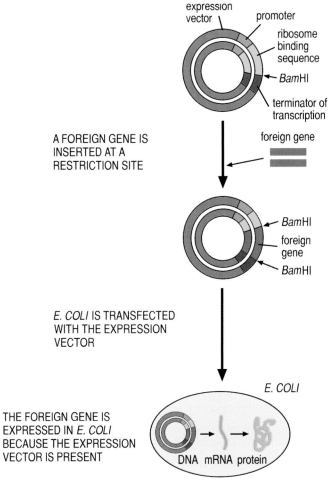

Figure 8.1 Recombinant DNA technology facilitates the production of human proteins in bacteria.

Tricks of the trade

Despite the fact that the principles of gene expression are fundamentally similar in bacteria (Figure 8.2) and humans, there are actually two basic differences which prevent human genes from being decoded by the bacterium without a great deal of prior manipulation. First, promoter sequences in human genes are not recognised by the bacterial transcription factors and RNA polymerase that regulate the production of messenger RNA molecules;

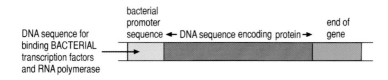

DNA sequence for binding BACTERIAL transcription factors and RNA polymerase

bacterial promoter sequence ← DNA sequence encoding protein → end of gene

1. TRANSCRIPTION FACTORS BIND TO PROMOTER REGION

transcription factors

2. RNA POLYMERASE RECOGNISES THIS AND STARTS TO MAKE MESSENGER RNA

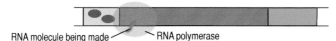

RNA molecule being made — RNA polymerase

3. RNA POLYMERASE MOVES ALONG THE DNA PRODUCING AN EVER-INCREASING LENGTH OF RNA

RNA polymerase

4. RNA POLYMERASE LEAVES THE DNA RELEASING THE RNA

new RNA molecule

5. RIBOSOME MAKES THE PROTEIN

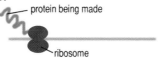

protein being made

ribosome

Figure 8.2 The production of a 'normal' bacterial protein. (Human genes lack the bacterial promoter sequence and encode RNA molecules that require splicing after step 4.)

this prevents human genetic information being transcribed in bacteria. Second, the genes of *all* multi-cellular organisms (humans included) have sections (called *introns*) which are not actually decoded. This means that the primary transcript has to be 'edited' ('spliced' is the word often used) to remove the introns and produce a mature RNA before it is used by the ribosomes to assemble the encoded protein. Bacteria do not have this system for splicing the messenger RNA and so are unable to process human gene messenger RNAs for use by their ribosomes.

These problems can be solved by isolating *mature* (i.e. spliced) RNA from the appropriate type of human cells and using a special viral enzyme (called *reverse transcriptase* which makes DNA from RNA) to produce a DNA copy (called cDNA) of this information. This cDNA now lacks introns because it was made from a template that carried the *spliced* information. The cDNA is then genetically engineered into a bacterial promoter sequence in a plasmid (Figure 8.3). When this genetically engineered version of the human gene is transformed into bacteria, the bacterial system recognises the bacterial promoter and produces the appropriate messenger RNA from the modified human DNA template. This message no longer needs to be spliced and the ribosome should therefore make the appropriate protein.

The feasibility of making a human protein in bacteria was demonstrated in the late 1970s in a heroic experiment: a small gene, 42 bases long, was synthesised chemically (a very tough job), inserted into a special plasmid carrying a bacterial promoter and, after much tweaking, a small amount of the encoded protein was made inside *E. coli*. That particular protein was of no great medical significance but, encouraged by this success, an even more demanding project was then undertaken. This time the intended product had a serious clinical significance: it was human growth hormone protein. The cDNA method was used and the product arrived just as the traditional source of growth hormone came into disrepute.

Into the market place

Situated at the base of the brain is a small pea-sized structure called the *pituitary gland*. Referred to as the 'master gland', it controls the body by releasing protein hormones into the bloodstream. These eventually bind to various cells throughout the body and direct them to perform numerous functions. One of the pituitary hormones regulates growth and a deficiency, though fairly rare in the population, results in stunted growth, the condition of *pituitary dwarfism*. In the 1950s a new therapy was developed for this condition: pituitary glands were removed from cadavers of people who willed their bodies for medical purposes and growth hormone extracted by a chemical process. When still young and growing, the patients with a growth hormone deficiency received injections of

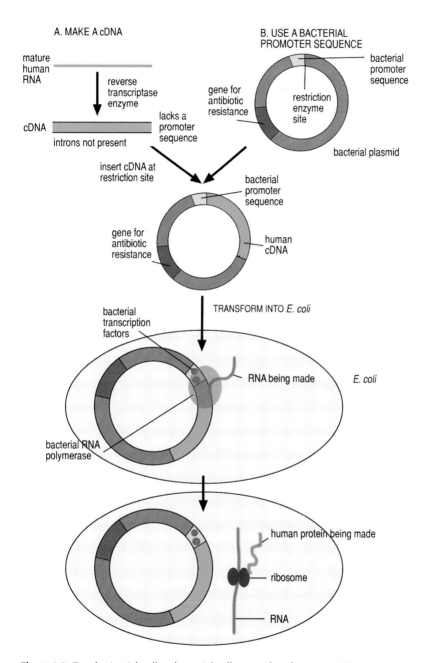

Figure 8.3 Two basic tricks allow bacterial cells to produce human proteins.

the protein at regular intervals and grew to a normal height. However, the hormone was extremely expensive because it was in such short supply.

In 1979 Genentech, one of the very first biotechnology companies, announced that they had succeeded in using a cDNA fused to a bacterial pro-moter sequence to produce human growth hormone in *E. coli*. At the same time as Genentech were testing their new product, three patients who had been treated with the pituitary extract back in the 1950s and 1960s suddenly developed Creutzfeldt-Jacob's disease (CJD, a brain degenerative disease anal-ogous to BSE in cattle); apparently a pituitary gland had been harvested as a source of growth hormone 20 years earlier from the cadaver of an individual who had been suffering from CJD. As CJD was not known at the time to be an infectious condition, the contaminated pituitary gland had been processed and the disease passed on via treatment with contaminated growth hormone.

The old procedure for producing the protein from cadavers was imme-diately banned. Luckily the recombinant version was just coming on stream and doctors switched their patients to its use. The new product was innocu-ous when tested on animals and had the desired therapeutic effect on humans. However, patients who were moved to the new product suffered from fevers and chills that had not occurred with the old version of the hormone. It was found that, although the protein was actually very pure, a small amount of membrane from the bacterial cells used to produce it was also present in the final product. Despite the fact that rats and mice tolerated this without effect, it had a noticeable and distressing side effect on humans. The problem was avoided by changing the purification procedures but it did highlight just how careful we must be when dealing with new technologies.

As large quantities of the hormone at last became available, other thera-peutic avenues were explored and growth hormone found a use in a number of conditions (such as in the healing of wounds and bone fractures) where rapid cell growth is required. It has even been used to reverse some of the effects of ageing and as an illegal performance-enhancing drug for athletes.

Encouraged by this success, the way was now open for the application of the same technology to other therapeutically important proteins. The next human protein to be made in *E. coli* was insulin. This is yet another example of a protein hormone. It is produced in the pancreas, transported around the body and regulates blood sugar levels by stimulating liver cells to take glucose from the blood and to convert it into glycogen – a storage form of sugar. A lack of insulin results in the glucose remaining in the bloodstream and giving rise to the condition called *diabetes*. Insulin replacement therapy started in the early 1920s when the hormone was first purified by chemical extraction from pig or ox pancreas and then injected into patients. There were, however, three problems:

- the insulin was not pure because it was contaminated with other proteins;
- the amino-acid sequences of porcine and bovine insulin are slightly different from the human form, resulting in some patients mounting an immune response against it;
- two pigs were required per person per week to maintain the insulin supply.

The human form of the protein was made in *E. coli* by isolating messenger RNA from a human pancreas and using it to make cDNA. This was spliced into an appropriate regulatory sequence in a plasmid and the bacteria obligingly produced the protein. As with human growth hormone, the producing cells were grown in huge numbers using enormous fermenters from which the cells were harvested, broken open and the target protein purified free from all other cellular debris before being extensively tested, bottled and sold. The human product now occupies a major part of the world market in this protein.

Novel products
The bacterial production of human growth hormone and insulin marked the beginning of a revolution in the way therapeutic proteins were manufactured, but they did not add to the medical armoury because these (or very similar) proteins were already on the market. Enthused by this success, biotechnologists now turned their attention to a group of extremely novel, rare and potentially powerful proteins called *interferons*. Discovered almost 25 years earlier, they are produced when a virus attacks a human or animal cell. These small proteins diffuse to nearby cells and block subsequent viral infections. Early analysis showed that they were also useful in inhibiting the growth of cancer cells. Unfortunately the quantity of interferon produced is so small that gallant efforts were required to purify enough even for a small-scale experiment to treat just a few patients.

Not surprisingly, the early biotechnology companies cloned interferon genes into *E. coli*, thus facilitating large-scale production of these precious proteins. Many research programmes and treatment regimes followed and a licence was granted for the sale of one of them as early as 1986. Interferons decrease the rate of growth of a number of the common cancers but are less effective against other more prevalent ones like those of the breast, lung and colon. Nevertheless, many interferon proteins are now being produced for use both in conjunction with other more traditional cancer therapies (e.g. chemotherapy) and for the treatment of virally related conditions like genital warts, multiple sclerosis and AIDS.

The pioneering gene cloners were greatly encouraged by the successful production of these human proteins in *E. coli*. However, their early successes were achieved with relatively small, uncomplicated proteins, exceptions rather than the rule in human cells. Success dropped off rapidly when attempts were made to apply the same technology to the production of bigger and more complex ones.

A sugar coating

When the amount of oxygen in the blood falls for a period of time, cells in the kidney respond by secreting a hormone into the blood. It travels to the bone marrow and causes extra red blood cells to be made so that more oxygen can be trapped in the lungs. This protein, which stimulates the production of erythrocytes (red blood cells), is called *erythropoietin* (EPO). It is frequently in the news as a banned substance used by some athletes artificially to enhance their performance.

Many patients with chronic kidney failure cannot make EPO and, as a consequence, are anaemic because of a lack of red blood cells. Giving regular blood transfusions has its own discomforts and inconveniences, not to mention cost and the possibility of mounting an immune reaction against repeated transfusions. Injecting the human protein into such patients could solve all of these problems but the quantities to be harvested from human urine were insufficient for the demand. EPO was therefore an ideal target for genetic engineers.

EPO is similar in size to the interferons, so it came as no surprise that *E. coli* churned this hormone out when the cDNA encoding was put through its production paces in the bacterium. The recombinant EPO bound to the correct type of cell in the test-tube and so, prior to being trialled in humans, it was injected into mice to assess its efficacy at increasing the number of red blood cells. This time a surprise was in store for the genetic engineers: the recombinant protein was quite useless at increasing the red blood count – yet another gentle reminder from Mother Nature that subtleties abound in biology!

Human cells are far more complex then bacterial ones. Not only do they contain lots more genes, but each cell is also divided internally into compartments. One of these is full of enzymes processing proteins for export. The exported proteins facilitate communication and co-ordination between the billions of cells making up the body. During the export process large and complex proteins are properly folded and sugar molecules are added to protein surfaces to alter their shape and/or enhance their strength and durability. Not surprisingly *E. coli*, which has no need to communicate with other cells, has no export department and thus no means of adding sugar to proteins. Some human proteins (the interferons amongst them) can perform well even without the sugar molecules – EPO cannot. It seems that, under normal circumstances, the thickness of the sugar coat determines how long the protein survives in the bloodstream. As these protein molecules age, they lose their sugar and are broken down. The EPO made in *E. coli* lacked the sugar coat completely and was therefore unstable when injected into experimental animals. The body degraded it before it had a chance to trigger an increase in the production of red blood cells.

New hosts

Despite the fact that *E. coli* is such a useful cell for producing many human proteins, the EPO story demonstrates its serious limitations. The inability to add sugars to proteins is one such problem; another is its inability to fold large complex proteins into their correct shape. For these reasons, shortly after their first successes in bacterial cells, scientists attempted to adapt the same technology to other cell types.

The first choice was the simple yeast cell because not only does it have primitive versions of all the major cellular compartments of human cells, but it also has many of the characteristics that made bacteria the choice for the original cloning system. Yeast grows rapidly as single cells, forms colonies on plates and has a plasmid, so allowing it to be genetically engineered more easily than complex organisms. Furthermore, yeast has a long history of safe use in the brewing industry. The first commercially successful recombinant vaccine was produced in yeast after *E. coli* was found to be unable to fold it properly. Before the advent of genetic engineering, the hepatitis B vaccine had been prepared by purifying the relevant protein from the blood of individuals who were suffering from the disease – a worrying thought when one considers that many patients who have the hepatitis B virus have also acquired the AIDS virus. Recombinant DNA technology enabled scientists to clone the viral gene encoding this protein, splice it into the regulatory sequence from a yeast gene and make the viral protein in yeast. Thanks to gene expression technology, a jab in the doctor's surgery is now much safer.

For a variety of reasons, other types of host cells for gene cloning and protein production were also developed. They included a number of bacteria and fungi as well as a virus that infects insect cells. Eventually it even became possible to use mammalian cells; this was especially demanding, as these cells by their very nature grow best inside bodies. However, certain types of human and other animal cells can be persuaded to grow in fermenters by providing them with an extremely complex liquid food supply. In fact, it was in this system that the elusive biologically active EPO was eventually made. The strategy with all of these host systems is the same: the gene of choice is spliced into a DNA molecule carrying the gene regulation sequences from the intended host cell together with a gene that the host cell has to have for growth under certain circumstances. This entire DNA molecule is then introduced into host cells and the event is selected for by transferring the host cells into a medium that only allows cells carrying the new DNA molecule to survive because they are the only ones able to grow. For example, one might culture the host cells in the presence of an antibiotic or other chemical, such that only those cells with the new DNA molecule carrying a resistance gene for the antibiotic or chemical can grow. Each one of the host cell systems has advantages and disadvantages in terms both of the technical procedures required to use it and of cost. It is therefore true to say that no host can be all things to all genes!

Pharm animals

The ideal host cell for producing proteins for therapeutic purposes would be a human or animal cell culture system growing in huge fermenters. Unfortunately such cell systems grow slowly and the production process is expensive, because extremely costly procedures have to be followed in order to prevent contamination of the cells or of the protein product with unwanted microbes. An alternative is to use genetic tricks to get the proteins produced *inside* an animal's cells. Ideally the protein should be produced in cells which will continually export it from the body, thus obviating the need to kill the animal to recover the product.

Milk supplies young mammals with all their food requirements for the first few weeks of life and therefore, along with many other goodies, milk contains proteins. These proteins are uniquely present in milk because the genes that encode them have regulatory sequences that are activated only in the breast cells that specialise in milk production (Figure 8.4). Genetic engineers can now take those regulatory sequences from the genes of a domestic animal like sheep, goat or cow and splice them onto a cDNA encoding a human protein. They use a micro-needle to inject this gene construct into an egg which, after *in vitro* fertilisation, is implanted into a surrogate mother. If the gene constructs successfully integrate into the chromosomes of the egg, the animal will produce the human protein in its milk when it is old enough to lactate. The procedure is extremely demanding and difficult to execute but such modified animals offer huge advantages over growing complex cells in fermenters because:

- there is a low capital investment and low operational costs, i.e. no need to buy and run expensive fermenters. Sheep eat grass, not expensive and complex growth medium;

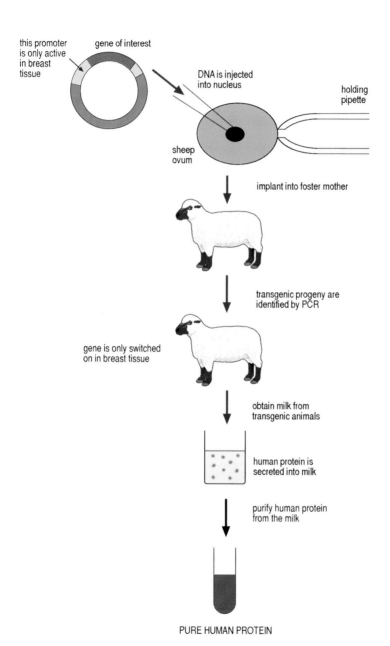

this promoter is only active in breast tissue

gene of interest

DNA is injected into nucleus

holding pipette

sheep ovum

implant into foster mother

transgenic progeny are identified by PCR

gene is only switched on in breast tissue

obtain milk from transgenic animals

human protein is secreted into milk

purify human protein from the milk

PURE HUMAN PROTEIN

Figure 8.4 Many therapeutically important human proteins are produced in the milk of genetically engineered sheep or goats. After Griffiths AJF, Gelbart WM, Miller JH, Lewontin RCW, Freeman H. Modern Genetic Analysis 1999.

- it provides ease of collection and purification. Milk can be collected using commercially available automated milk systems and, as milk contains but a few sheep proteins, purification is reasonably straightforward. In their raw state, products made in fermenters have many more contaminating proteins, while harvesting is arduous and the purification procedures much more cumbersome;
- production capacities are high. Throughout a 5-month lactation period, sheep can produce 2–3 litres of milk per day containing 1g or more of the desired therapeutic protein per litre;
- breeding more individuals carrying the same gene construct can readily increase the total output.

The proteins produced in milk have had their sugar coating added and are folded properly so they have the appropriate biological functions. Therapeutically useful proteins made in this way include: tissue plasminogen activator, an enzyme for breaking up blood clots causing heart attacks, α_1-antitrypsin, a protein helping to prevent the lung damage that culminates in emphysema, and the blood clotting factors VIII and IX which are used to treat bleeding diseases called haemophilia. With the blood clotting factors, hitherto isolated from blood supplied by donors, production in animals removes all risk of the products being contaminated with the AIDS virus which, tragically, has happened in a few cases in the past.

So far, not many transgenic animals have been bred because there is a high failure rate and each batch of animals takes many months to test. Nevertheless, the first of these products is now entering clinical trials and, as we shall see, this technology has been given a huge boost by the successful cloning of Dolly – but much more about this in Chapter 9.

The application of recombinant DNA technology has come a long way since the triumph of producing the first human protein in bacteria in 1977. *E. coli* is still the organism of choice whenever possible and indeed the list of proteins produced by *E. coli* is impressive (Table 8.1). Other protein production systems are used for a variety of proteins but mammalian cells in fermenters or sheep cells in udders are extremely good at producing large, complex, processed proteins and are often the systems of choice for producing therapeutic proteins.

Recombinant DNA technology is particularly good at producing proteins whose therapeutic value depends on their availability in the bloodstream. However, there is not a lot that this type of therapy can do for thousands of other conditions in which the defective proteins are to be found *inside* the cells; these include sickle cell anaemia, cystic fibrosis, Huntington's chorea, many cancers and a multitude of other conditions. Recombi-

Table 8.1 Proteins produced in bacterial cells

Protein	Quantity of the *human* protein made in the bacterial cells (As percentage of *all* of the proteins)
Insulin	20
Growth hormone	5
Interleukin 2	10
β interferon	25
α_1-antitrypsin	15
Human tumour necrosis factor	15

nant DNA technology could, of course, be used to make lots of the missing protein in a fermenter but it would be impossible to get the product into the appropriate cells of the patient at the proper time and in the correct amounts. Faced with the fact that they cannot transplant functional proteins into cells, genetic engineers are attempting to rectify these conditions by a much subtler scheme. It is so powerful that, if successful, it would be able to cure not only those scourges listed a few lines ago but make even the exciting technologies described in this chapter very rapidly redundant.

9

Wonderful cures

'Go and stand out in the rain for two hours without a hat or coat', said the doctor to the young man complaining of joint pains, shivering, coughing and sneezing. 'But I shall get pneumonia' complained the patient. 'Good', replied the doctor. 'I can cure pneumonia but I cannot cure the 'flu.'

There is a lot of truth in this old joke. Antibiotics are extremely effective at fighting pneumonia caused by bacteria because they kill the invaders as they try to grow *outside* the cells at the site of infection. The 'flu virus, on the other hand, must, in order to reproduce, introduce its genetic material *inside* a living cell. Inert outside the cells and 'alive' inside, viruses are almost impossible to combat without compromising the well-being of the cells they hijack.

Viruses are in many respects molecular syringes that specialise in injecting their genes into living cells. That rang a bell for genetic engineers involved with human diseases caused by genetic defects: could viruses somehow be used to smuggle the normal (non-mutant) form of the gene into affected cells? Scientists are now attempting to do just that by replacing some of the viral DNA with normal human genes. They then try to persuade the viruses to introduce these genes into cells that carry serious mutations. If they succeed, they will have achieved extraordinary cures for diseases arising when mutations result in defective proteins.

The problem

The vast majority of diseases caused by defective proteins involve proteins that are found *inside* cells. Sickle cell anaemia patients have defective haemoglobin proteins in their red blood cells; cystic fibrosis occurs because the protein that allows salt transport is not inserted into the cellular membrane. Cancer cells contain abnormal proteins responsible for uncontrolled cell division. There is no point in using expression systems to produce normal versions of these proteins in fermenters because there is no way of getting them into the cells in order to carry out their functions. Abnormal proteins are produced in cells that carry mutated genes. Cells carrying normal genes produce normal proteins; the obvious solution therefore is to arrange somehow to *repair or replace* the defective gene!

Indeed, as only a subset of genes is active in any given cell type, the problem boils down to repairing or replacing the defective gene in only those cells in which the dysfunctional protein is being made. Thus, in the case of sickle cell anaemia, it would be necessary to repair the gene mutation only in premature red blood cells; for cystic fibrosis, repair would be required for the mutated DNA only in lung cells and pancreatic cells but not in other types of cell. The logic of this argument is flawless, the technical requirements immense: producing copies of normal genes is easy enough – simply clone them from the cells of an unaffected individual. Getting these genes into all (or even a high percentage of) the target cells, in such a way that they are correctly regulated and the normal protein produced, is a very much more difficult task.

Attack by the living dead

There is an enormous variety of viruses and they attack every organism from bacteria to humans. Their modus operandi is to find a suitable organism to infect, gain entry to an appropriate cell (e.g. ciliated cells in the windpipe for a 'flu virus) and hijack its resources to make viral proteins and viral genetic material. Viruses are so small they cannot be seen using even powerful light microscopes; exploring their physical structures depends on sophisticated analyses using electron microscopes and X-ray diffraction procedures. Viruses show a very wide range of sizes and shapes (Figure 9.1) but generally comprise a small amount of genetic material (usually DNA but sometimes RNA) wrapped up in a coat made of protein; some are further enclosed in a

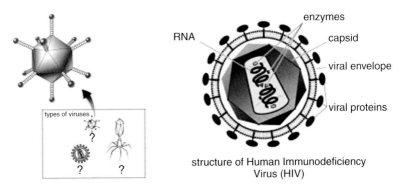

Figure 9.1 Despite the simplicity of their structure, viruses come in a variety of shapes and sizes. Reproduced courtesy of the United States Department of Agriculture.

droplet of fat (called a *lipid membrane*). Books have been written on the various strategies that different viruses use first to gain entry to a living cell and then to replicate their genetic material.

Attack by the 'flu virus is depicted in Figure 9.2 (a). Protein 'spikes' embedded in the lipid layer of the virus bind to a protein which protrudes

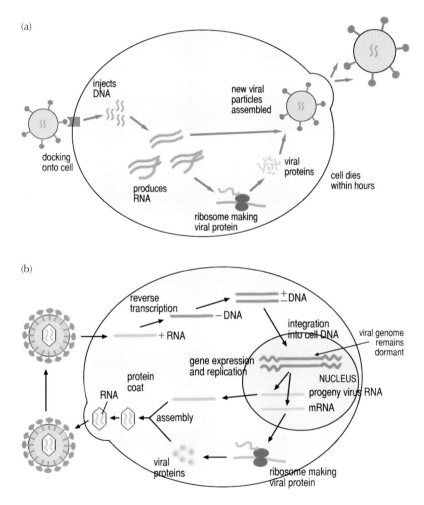

Figure 9.2 Different viruses attack different types of cells and hijack their metabolism.
     (a) The 'flu virus attacks cells in the upper respiratory tract and immediately hijacks it to destruction.
     (b) The AIDS virus invades cells in the immune system and integrates its DNA into the host cell's chromosome. It can be there for years.

from the membrane of a special type of cell in the respiratory tract of the victim. These cells, present in the nose, throat and windpipe, possess small short hair-like structures called *cilia* which beat rhythmically to spread protective mucus over the airways and protect them from microbial invasions. After successfully docking onto the target protein, the virus is taken into the cell where it releases its own genetic material and sets about forcing its host to produce new viruses. These hijacking activities result eventually in the death of the ciliated cells, so exposing the layer of cells lying beneath. Luckily these do not have the docking site required by the virus and so they escape invasion. The many millions of viral particles present in the respiratory tract trigger the body's immune system and it starts to remove the invader. Some 7–10 days after the first attack, the body's defences have defeated the invader and the ciliated cells can safely regenerate. Barring the occurrence of secondary infections by other bacteria or viruses, the patient makes a full recovery within 14 days. It is to prevent *bacterial* pneumonia that doctors sometimes prescribe antibiotics for 'flu victims. *Antibiotics do not kill viruses*; hence their uselessness for treating colds and 'flu.

Unfortunately the body's defensive capabilities are not always so effective at removing the invader. Many viral diseases, including hepatitis and polio, can leave permanent effects on people they attack; others, such as yellow fever, rabies and HIV frequently kill the unfortunate victim.

HIV – a particularly nasty example of the living dead

On the face of it, the virus that causes AIDS in humans shares features common to lots of other less harmful viruses. Its devastating impact arises because of the particular type of cell that it hijacks in order to replicate. As patients recover from the 'flu, their immune system can remove the virus and thus allow the body to regenerate the cells which have been destroyed. The AIDS virus, however, attacks one of the key cell types in the immune system and eventually wipes it out. Shorn of a working immune system, the victim usually succumbs not to the viral infection itself but to secondary infections which the body cannot fight in the normal way. It is the immune deficiency which is the most virulent aspect of the HIV infection and which gives this dreadful disease its acronym *AIDS* (**A**cquired **I**mmuno**d**eficiency **S**yndrome).

The AIDS virus can infect a person only if it is transmitted directly into the bloodstream, because it is there that it can find the cells it needs to invade in order to survive. The virus has outer-coat proteins significantly different in structure from those of the 'flu virus; they enable the infective agent to dock primarily onto cells of the immune system. Unlike the 'flu virus that immediately sets about hijacking the metabolism of the cell in order to replicate, the AIDS virus carries its own special enzyme which allows its genetic material to become incorporated into the host's own DNA (Figure 9.2 (b)). Those viral genes sit there for a long period, often years, until that immune cell is called upon to fight infection by some other invader. Immune cells frequently divide as part of their mechanism to repulse attacking invaders: it is when the immune cell divides that the viral genes become activated and it is only then that the cell is hijacked. The viral particles are assembled and the host cell releases mature virus particles into the bloodstream where they infect other cells. It is then that the patient gets full-blown AIDS. Spreading from cell to cell, the virus destroys them as it goes so that the victim is no longer able to mount immune responses to *any* infection. Eventually, one or other infection causes sufficient damage to kill the patient.

These two examples illustrate a number of critically important points about viruses:

- different viruses attack different cell types;
- some can integrate themselves into the DNA of the cell they infect;
- when they replicate, they eventually kill the cell that they have invaded.

Revenge on the living dead

For millions of years viruses have been injecting their genetic material into specific cells of susceptible organisms. If it were possible to exploit their docking and injection specificity, while genetically engineering them so that some of their own genes were replaced with human ones, it might be possible to change these scourges of humanity into therapeutic tools.

As we have seen, viral particles responsible for AIDS dock onto cells of the immune system and release their genetic material. It is possible to use recombinant DNA technology to remove the viral genes responsible for hijacking the cells' metabolism but leave intact those genes that encode the viral coat protein. The removed viral genes are then replaced with human genes encoding useful proteins like interleukins, which co-ordinate the activities of the various specialised cells constituting the human immune system. In theory, such re-engineered viruses could be used to deliver life-saving gene therapy to the very cells the unmodified virus would normally attack and kill. Similarly, it should be possible to reconstruct a 'flu virus so that it carries a normal gene for the cystic fibrosis protein in place of the genes that have been removed from the viral genome. Were such a virus sprayed into the respiratory tract of a cystic fibrosis patient, it would attach and release its genetic cargo as usual. But instead of destroying them, the modified virus would deliver a gene allowing the cells to produce the normal version of the cystic fibrosis protein and thus alleviate the disease. Such an approach is possible because cystic fibrosis is a recessive condition in which the patient has *two bad genes* for the protein, one from each parent. It would be necessary to provide just *one good gene* for the condition to be rectified.

Turning harmful viruses into potential gene therapy tools is a relatively straightforward task (Figure 9.3). Recombinant DNA technology is used to remove all the dangerous viral genes from the viral genome. This leaves the genes that encode the coat proteins and the extreme right and left hand ends of the genome. The human gene (or genes) of choice is then inserted into this DNA molecule and it is transformed into a mammalian cell in a test tube. The cells produce multiple copies of this 'viral-human' DNA molecule (in a normal situation these would all simply be viral genomes). The cells also produce coat proteins encoded by the few viral genes left in the DNA molecule. The latter wrap the DNA into a protein coat that is the same as the virus normally has – except now there is no virus but just human DNA wrapped up in a viral envelope. After harvesting and purification, these engineered particles are ready for use. They are still able to dock onto the target cells but, as the genetic material that they deliver does not contain the complete viral genome, they cannot subvert the host's cellular metabolism. Instead, they deliver a normal human gene which the cell could use as its own for producing the encoded protein (Figure 9.3).

Developing and testing the safety of viral vectors is an expensive business and, so far, most research has concentrated on a handful of viruses with a narrow range of characteristics. Viruses related to AIDS that can integrate their genetic cargo into the chromosome of target cells, and so theoretically require only one treatment, are extremely desirable as gene vectors. However, they cannot carry much DNA and are able to infect only *actively* dividing cells. This means that such viral vectors are limited to delivering *small* genes to *dividing* target cells. On the other hand, vectors related to the 'flu virus can incorporate DNA inserts and deliver their genetic cargo to *non-dividing* cells but they do not promote *integration* into the target cell's chromosomes. It is therefore necessary to apply repeated gene treatments which, as we shall see, restricts their applicability. Non-viral vectors, such as lipid (fat) bubbles carrying DNA within them, have also been developed but are difficult to target to specific cell types. In fact, current thinking favours the development of new hybrid vectors carrying various bits and pieces from a range of different vectors, e.g. a hybrid combining the carrying capacity of the 'flu virus with the ability of the AIDS-type viruses to integrate their genetic cargo into the cells. Even then there are problems and the experts in this field agree that considerable effort on vector design is still needed.

Assembling the different tools for gene therapy, however, is just the beginning of this procedure. The major stumbling blocks of this technology lie in the delivery and expression of the genes.

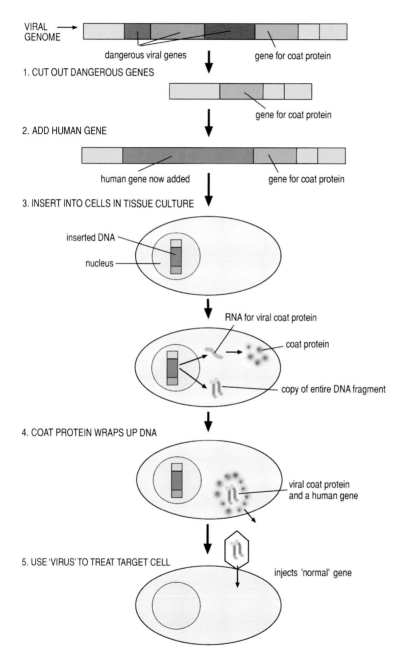

Figure 9.3 A virus can be genetically engineered into a gene delivery system.

Out of body experiences

Gene therapy is a novel concept and, not unreasonably, is subject to stringent regulation. Injecting re-engineered viruses into humans is not to be undertaken lightly. It is for this reason that the first attempts were performed with cells which could be removed from the donor, treated with the new gene, checked for any unforeseen problems and then re-introduced into the patient's body. The candidate diseases chosen in the early days of human gene therapy were ones caused by inherited single gene defects because such diseases had been analysed for years, their mechanisms were well understood and the relevant normal genes were available.

The first case of gene therapy was approved in 1990. Scientists used a re-engineered AIDS type virus to insert a normal human gene into cells from the immune system of a 4-year-old girl suffering from a rare inherited immune deficiency disease caused by the lack of a well-characterised enzyme. Individuals with this disease have no properly functioning immune system and are forced to live in a special sterile environment in order to survive (Figure 9.4). The virus delivered the gene into the cells, which were grown with special nutrients in the laboratory until there were a thousand million cells; then they were transfused back into the patient. This approach has been tested a number of times; there is a consensus of opinion that the procedure itself is safe and that, in some cases at least, the patients show some improvement in their condition.

There was a similar result with a much more difficult type of therapy concerning a rare inherited condition called familial hypercholesterolaemia, resulting at all times in dangerously high levels of cholesterol in the blood. The cause is the lack of a protein which helps to harvest blood-borne cholesterol into the liver. A segment of the patient's liver was removed, cells were grown in a test tube for a while and infected with a virus intended to insert the gene for the normal protein. The modified cells were injected into the patient whose blood carried them to the liver where they re-colonised the damaged organ. The patient did show some improvement but, as with the other experiments, after a short while and for no apparent reason the 'new' gene switched off. Nobody so far has been permanently cured of his or her dysfunctional proteins.

When it was first conceived, gene therapy was considered to be the answer to inherited single gene defects. Now the emphasis has shifted from treating rare inherited disorders to the much more common acquired disorders like AIDS, heart disease and cancer. Indeed, the greatest efforts are directed towards cancer therapy. And there are millions of good reasons for that decision. Biotechnology and pharmaceutical companies investing huge amounts of money must see a return on their investment: no perceived

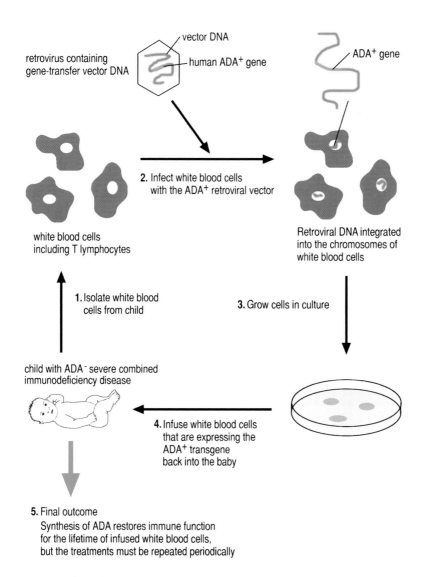

Figure 9.4 The earliest gene therapies required cells to be removed from the patient, engineered and then reinjected into the patient.

return, no investment. Commercial companies are not charities – they exist only to make profits for their shareholders and the cancer market is far bigger than the market for any of the *inherited* defects. Furthermore, clinical trials are rigorously regulated and obtaining approval for new therapies is very difficult. It is relatively easier to obtain permission for clinical trials on terminally ill patients than for conditions like AIDS for which reasonably effective therapies already exist. The approach being used in cancer therapy is instructive in that the genetic engineers are not trying to repair the defective gene in the cancer cells, but rather to trigger the immune system to attack the tumour. Nevertheless, it still uses an out-of-body approach to the therapy.

Killing cancer cells

Our bodies contain so many billions of individual cells that it is not too surprising to hear that mutated cells frequently arise capable of producing a tumour (or cancer). Under normal circumstances, our body's immune system detects that tumour cells are not 'normal' cells and sets about destroying them before they can do any harm. But sometimes the tumour cell escapes, either because it goes undetected by the immune system or because it divides faster than the immune system can cope with it. A medical treatment that either unmasked the tumour cells or augmented the immune response would aid in the control of tumour development.

Gene therapy offers hope for both approaches. In one, cells are removed from the tumour, grown in tissue culture in the laboratory and then transformed by adding a gene which alters their surface properties and makes them more easily detected by the immune system (Figure 9.5 (B)). Finally, the modified cells are treated with X-rays to prevent them dividing and injected back into the original patient. The effect is to activate the immune system to that type of tumour cells in a way analogous to a vaccination. Once the immune system has been primed, the defences are much more aggressive in their attack on cancer cells *of that type*. An alternative therapy is to remove immune system cells from the patient, grow them in tissue culture and inject them with a gene which encodes an extra protein. Such proteins are able either to attack tumour cells directly or to act as an alarm to attract other immune system cells to deliver the *coup de grace* (Figure 9.5 (A)). These procedures for treating cancer have worked very impressively in mice and a number of clinical trials are currently attempting to do the same for humans. The outcome is being awaited with great interest.

The Holy Grail

The theory of gene therapy is elegantly simple; the practice is fraught with technical limitations. All the examples outlined above require removing the target cell from the body, applying gene therapy and returning the cells to the patient. The Holy Grail of gene therapy is to develop systems which can be administered directly to the patient. It would replace cumbersome tissue

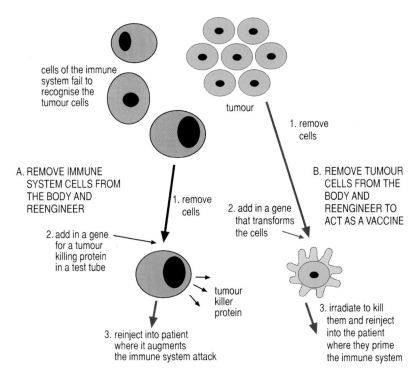

cells of the immune system fail to recognise the tumour cells

tumour

1. remove cells

A. REMOVE IMMUNE SYSTEM CELLS FROM THE BODY AND REENGINEER

1. remove cells

2. add in a gene that transforms the cells

B. REMOVE TUMOUR CELLS FROM THE BODY AND REENGINEER TO ACT AS A VACCINE

2. add in a gene for a tumour killing protein in a test tube

tumour killer protein

3. irradiate to kill them and reinject into the patient where they prime the immune system

3. reinject into patient where it augments the immune system attack

Figure 9.5 Cancer treatments using gene therapy are focused on augmenting the immune system's ability to recognise and destroy tumour cells.

culture procedures with routine drug administration and allow the application of this technology to *all* cell types, including those that cannot be removed from the body.

A direct approach must be used in any attempt to treat the lungs of cystic fibrosis patients. 'Flu type viruses, re-engineered to carry the normal gene for the defective protein, have already been tested in the clinic. They were inhaled into the respiratory tract but a combination of the mucus there and the fact that this viral system does not allow the gene to integrate into the chromosome meant that multiple treatments were required. Unfortunately, the immune system attacked the virus and caused sufficient problems in clinical trials for this procedure to be abandoned. This is always a problem with repeated viral treatments because such procedures offer the patient's immune system an opportunity to work up a response to the vector, which would not be the case if only one treatment were required. These viral vectors (i.e. gene carriers) need further genetic modification to disguise them and prevent their recognition as foreign by the immune system. Trials of other systems aimed at the disease are, however, continuing.

Other problems arise when developing viral vectors:

- it is extremely difficult to get blood-borne vectors into the brain because it is protected by a barrier against infections;
- other vectors bind to a number of different cell types as they circulate through the body – ideally they would target one only;
- the insertion of the gene into body (somatic) cells is acceptable but, once in the bloodstream, the accidental insertion of a new gene into egg or sperm cells might have far-reaching consequences for the next generation.

In short, the practice has a long way to go before it catches up with the theory!

Even if an ideal vector were to be produced tomorrow, able to carry genes of any size, deliver to the precise target cell with unerring accuracy and get its genetic cargo safely integrated into a chromosome, the real problem would still have to be tackled. The sad fact is that even when genes are successfully targeted into cells, they work well for a while and the normal proteins are produced for days, maybe even weeks. But then, suddenly and inexplicably, they switch off and do not come on again. That is completely unexpected and nobody knows why. It is a serious problem which must be solved before effective gene therapy will ever become a reality in the clinic.

There is no doubt that many of the hopes pinned on gene therapy have been a little premature and the basic research laboratories will have to work hard to iron out the major problems besetting the technology. Equally certain is the fact that they *will* be ironed out – many millions of pounds are being spent per year on these problems and the long awaited answers are beginning to appear. Moreover, a technology capable of *completely eliminating* defective genes from the population has been thrust into the realms of definite possibility by the birth of a sheep called Dolly!

10

Amazing ambitions

Dolly entered history as the first animal clone. Yet, in truth, the epoch making aspect of her birth was the fact that she heralded an entirely new approach to the genetic manipulation of complex animals. Scientists had previously struggled to introduce even one gene into the cells of such an animal, but the birth of Dolly opened the door to the manipulation of entire genomes and with it the possibility of producing herds of genetically reprogrammed animals, creating a race of perfect human beings or resurrecting extinct species. Far-fetched perhaps, but actually far from being science fiction. The first re-programmed animals have already been produced; the conscious design of humans was the subject of a recently convened scientific conference and a project aimed at resurrecting an extinct species has just been initiated.

Re-programming animals

Molecular biologists have spent a good deal of time working out how genes are switched on and off in different cell types and how cells ensure that proteins are produced when and where they are required. This has allowed them to manipulate the switch sequences so that they can be added to genes not normally containing them in order to express their encoded proteins in a desired fashion. In Chapter 8 we saw that it is possible to re-programme the milk-producing cells of sheep to produce a human protein. Scientists used genetic manipulation to remove a genetic switch from a sheep gene, added it to a human one and put the modified gene into udder cells. In theory it should be possible to use a virus to undertake gene therapy on the sheep's udder but in practice this is fraught with difficulties. The gene was therefore inserted into a fertilised egg which was allowed to develop into a living sheep carrying the 'new' gene in every cell in her body, including the reproductive cells she could thus pass this gene on to the next generation. This ensured the continued production of the precious protein into the future without having to resort to further cloning. Furthermore, as an udder cell-specific switch activated the human gene, these animals produced the human protein only in their milk.

Prior to Dolly, this procedure required the use of a microscopic syringe to inject the gene into individual fertilised eggs. Four-fifths of the embryos did not survive the injection or failed to develop for other reasons. Of the survivors, only 1% incorporated the injected gene into all their cells and successfully expressed the encoded protein in the milk. Even then, these genes integrated at random in the DNA of the fertilised egg and, as some sections of chromosomes inhibit the activation of genes in a non-specific fashion, the level of gene expression – and hence human protein production – varied considerably from one animal to another (Figure 10.1A). Thus, the limiting factor in the production of transgenic animals was manually getting DNA into individual fertilised eggs. It is relatively easy, on the other hand, to introduce foreign genes into ordinary cells growing in tissue culture in the presence of certain chemicals simply by flooding thousands of them with the chosen DNA. The problem then, of course, is that – even when some of these cells have acquired the desired gene – they cannot be grown into animals because they are not fertilised eggs!

These were the problems that Dolly solved and they were very tough ones. The new technology allows novel (human or other) gene combinations to be introduced into easily manipulatable tissue cultures of 'normal' cells, removal of their nuclei (which now carry the extra gene) and injection into eggs from which the nucleus has been surgically removed. This eliminates the hit-and-miss of the gene injection method because nuclei are removed only from cells known to be carrying the human gene. Inserting gene constructs into enucleated eggs and growing the early embryos in the laboratory enables the stability of the new gene to be confirmed before placing the egg in the uterus of a surrogate mother. One year after the birth of Dolly, the same team of scientists used this precise procedure to produce Polly; she is a lamb carrying the human gene for a protein required by haemophiliacs to assist blood clotting. Nuclear transfer technology has now been successfully applied to cattle, mice and goats as well as sheep. It is clearly the method of choice for creating transgenic animals to make human proteins for pharmacological use (Figure 10.1B).

Researchers are now investigating the possibility of applying more extensive genetic manipulation to tissue cultured cells prior to using them in nuclear transfer. The objective is to re-programme animal development so that their organs lack the biological molecules making those organs unsuitable for transplantation into humans. Likewise, they wish to introduce multiple genes to produce large changes in the genetic programmes of the recipients. The speed with which animal cloning and animal re-programming have been thrust upon the scientific community has been nothing short of stunning, ensuring that the application of nuclear transfer technology to human cells is now firmly on the scientific agenda.

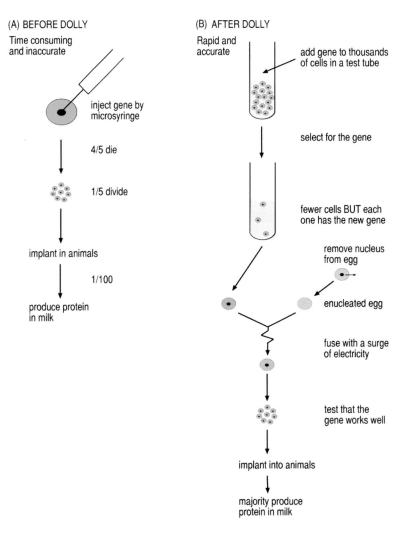

Figure 10.1 (A, B) Nuclear transplantation techniques facilitate the production of transgenic animals.

A rib too far

Whatever the perceived advantages of animal cloning technology, the vast majority of reproductive biologists have no intention of attempting to produce a human clone using the precise procedure for the creation of

Dolly. In fact, in many countries it is illegal even to try. There are several reasons for this, some technical, some practical and others ethical.

Dolly was not the product of a one-off experiment but the culmination of many years of hard work and frustration. Even the final successful experiment was a heroic piece of research, requiring the harvesting of hundreds of sheep eggs with the nucleus painstakingly removed individually from each one. Similarly, hundreds of adult udder cells had also individually to be manipulated under a microscope and a thin glass tube used carefully to withdraw the nucleus without damage for injection into the enucleated eggs. Of the 277 attempts in that experiment alone, only 27 embryos developed normally for the first week and were implanted in surrogate ewes. Half of those were lost in the last two-thirds of development, compared with a 6% loss in normal procreation. Some of the lamb foetuses were unusually large: the development of their hearts, livers and kidneys were profoundly changed. In addition to Dolly who did survive, other lambs were born live but died soon after birth. Even if the technique could be applied to humans tomorrow, the practicalities of harvesting hundreds of eggs and the mental trauma involved in subjecting a group of surrogate mothers to a pregnancy which stood a very high chance of failure would have serious ethical problems for everyone involved.

Were such a baby to be born, it might well be at risk of developing all manner of diseases. Remember that cells accumulate mutations during an individual's lifetime (Chapter 4); if a skin cell (which might be quite old) were used to generate a new baby, it would carry on where the old cell stopped. That baby might age prematurely or be prone to cancer. It might be argued that by ironing out these problems in animals, the way is made clear for using the techniques in humans. However, given the many differences in the reproductive physiology and embryology between different mammal species, it would be neither possible nor proper to use humans without a lot of human experimentation both in the laboratory and using live people. The standard medical view that new treatments of this nature should be sanctioned only when the potential benefits outweigh the risks to both baby and surrogate mother is a powerful and so far successful argument in delaying the legal acceptance of such techniques.

Suppose, however, that such a technology were to be used on rare occasions: what manner of psychological problems might both parents and offspring have to deal with? Identical twins are clones of one another but they differ in significant ways in terms of high brain functions. Furthermore, they know they have been born of parents in the normal way. At least 50% of our personality is thought to be environmentally determined, so a replica baby would almost certainly have its own distinctive personality. Would the sole parent be disappointed in the child if it did not turn out as expected?

What about social and genetic parentage? This technology requires an egg donor, a nucleus donor and someone to gestate the 'clone'. Who would be the real mother? If the female partner donated the nucleus – who would the father be? Indeed, *would* there be a father? Socially it would be the male partner (if there were one): genetically it would be the female partner's father who, incidentally, would also be the baby's social grandfather! Should the child be told about the confusing relationships? What might he or she think about having such a bizarre origin?

It is not difficult to see why the public appears disinclined to contemplate the development of this technology and why experimentation has been barred in many countries. There are, nevertheless, those in society (scientists amongst them) who argue that there are no scientific grounds for banning human cloning using the genetic information from adult cells. Rather, they argue, it should be legalised and closely regulated because:

- it would pose no threat to general human health;
- steps could be taken to minimise biological problems; and
- it could help a small number of individuals who are able to have children in no other way.

Indeed, the aptly named maverick scientist Dr. Richard Seed wishes to establish a number of profitable human cloning clinics with an initial objective of cloning his wife. Unlike Eve who, in the book of Genesis, became a unique and complementary company to Adam from whose rib she had been formed, the product of such an experiment would be a replica of an individual who was decades older than the clone. In the minds of many people, the psychological problems this sort of relationship would generate preclude such social engineering. In short, for the vast majority of individuals, cloning humans from cells taken from an adult is quite simply a rib too far!

Infertility treatment

However, the production of human clones using foetal cells is being actively pursued. For many couples, *in vitro* fertilisation is the best chance of having a baby. It requires the female to be given months of hormone treatment to increase the number of eggs she produces. These are then harvested and mixed with her partner's sperm to achieve fertilisation. Fertilised eggs are then grown for a few days in the laboratory until the cell divides and starts to form a ball of tissue. The most promising early embryos are inserted in the mother's womb where one or more can implant and go on to develop into a full-term baby (Figure 10.2 (A)).

Advances in this treatment have increased the success rate but couples often have to undergo more than one round of the treatment before it works.

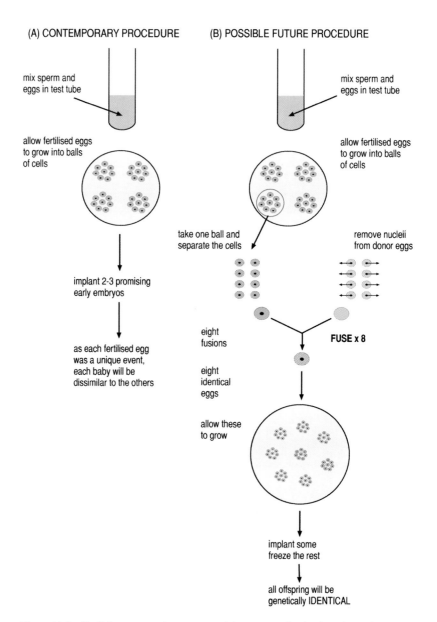

Figure 10.2 (A, B) It may soon become possible to treat infertility by a limited type of cloning that uses cells arising from a fertilised egg.

Super-ovulation and egg harvesting are not particularly pleasant experiences and any developments that cut down the number required are to be desired. It is here that the use of nuclear transplant technology can offer an alternative route. If a fertilised egg were to be allowed to grow in the laboratory through a few cell divisions, each cell could be used as the source of genetic material for the production of a baby. Nuclei could be transferred into enucleated donated eggs to produce 'new' embryos, so avoiding the need for multiple super-ovulation and egg harvesting procedures. A development like this is less contentious than the use of genetic information from adult cells because the cells are young and are less likely to have accumulated mutations; it removes the need for multiple hormone treatments to induce egg production and the donor is, in fact, a cell rather than a unique individual. But it is not without its downside, as it still requires the destruction of the original embryo and there is also the possibility of identical twins being born many years apart if frozen amplified embryos are used to produce separate pregnancies (Figure 10.2 (B)).

Tissue engineering

Nearly 160,000 patients die annually as they await an organ from a donor. Matching donors to recipients is a heartbreaking experience for all concerned and far too many lives are spent in hopeless anticipation waiting for a life-saving organ. Might the bottleneck be avoided?

Nuclear transplantation research seeks to understand how a *single cell* can divide, with the progeny cells differentiating to produce an embryo (and eventually a baby) consisting of many *different cell types*. We already know that the very early embryo produces 'stem cells' which are capable of differentiating into any type of cell. Studies are already under way to work out the various chemical factors directing stem cells to form specific tissues and organs. In the near future it should become possible to treat patients needing a transplant by injecting the genetic material from their cells into enucleated animal eggs and directing them to produce *one specific tissue or organ* rather than a whole embryo (Figure 10.3). Such tissue engineering has bone marrow as its first objective and there is at least one US biotechnology company already actively pursuing this objective. Eventually, human organs may be produced on demand. Once more, there are ethical problems because some people reject the idea of potential embryos not being allowed to develop into individuals. However, future research will no doubt reveal how eggs unmask the genetic information in the transplanted nucleus so that it can be used to make a whole individual. It will then become possible simply to unmask the genetic information of *any human or other cell* and get it to produce a specific tissue without going through the step of nuclear transplantation. Dolly has opened the door to all sorts of new technologies.

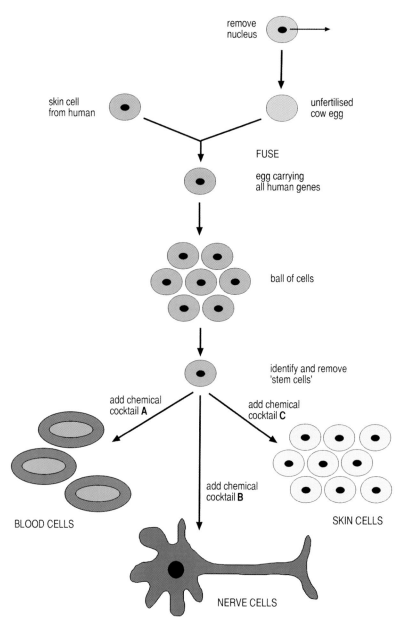

Figure 10.3 Nuclear transplantation and the isolation of 'stem cells' may facilitate the production of specialised cells, tissues and organs.

The ultimate cure

We all have two basic types of cell. Those that make up the body are the *somatic cells*, while eggs and sperm are the *reproductive cells*. As we saw in Chapter 9, gene therapy is currently aimed at the somatic cells but ironically, as it becomes more successful, it starts to exacerbate the healthcare problem it originally set out to solve. At present, individuals suffering from serious genetic diseases frequently fail to reproduce – either because of an infertility problem or premature death. If, as one hopes, such individuals are treated successfully, many will produce offspring. Unfortunately, the defective genes will still be present in their reproductive cells and their offspring may inherit the disease. An unfortunate consequence of somatic gene therapy may be an increase in the number of people in the population actually getting the disease. Additional health resources would be needed to treat them and a vicious upward spiral commences. The obvious answer to this problem is to use gene therapy on the reproductive cells but the chances of doing so are, for the moment, small. A better idea would be to correct the defect in a fertilised egg before it is introduced into the mother's womb but we have seen earlier in this chapter that reprogramming fertilised eggs is very difficult and unlikely to be successful (Figure 10.1 (A)). Dolly nevertheless demonstrated that *any* cell can be developed into a complete animal, so the answer may be to use nuclear transfer technology to remove defective genes from the population. When gene-cloning procedures become rather more refined it should be possible to take cells from an early embryo and add the 'normal' gene. If these cells were the nuclear donors for enucleated eggs, embryos would be produced carrying the normal gene in all their cells, including their reproductive ones, but this creates yet another ethical dilemma.

When people choose to have somatic cell gene therapy, they accept, as for any medical procedure, a chance that something could go wrong and that, if it does, they must live with the consequences. However, if germ-line gene therapy goes wrong, many individuals, perhaps for generations, will have to live with the consequences of an act over which they had no control. Does one have a right to take such risks with future generations? Not surprisingly, human germ-line gene therapy is everywhere illegal and, for the moment at any rate, there are no procedures available with which to do it.

Yet many scientists feel it is only a matter of time before it is attempted. One of the major reasons is the belief that it will be easier to tweak genes in a single cell in tissue culture than to inject them into millions of cells using somatic gene therapy. At a recent conference convened to discuss the scientific and ethical aspects of this technology, even scientists who were strongly opposed to the idea of germ-line gene therapy felt that it was only a matter of 10 years or so before the pressure to attempt it became irresistible. They felt the first attempts would involve repairing severe genetic defects such as Tay Sachs disease which, we noted in Chapter 7, is a very nasty, recessively inherited genetic condition, causing the brain to degenerate in early life. If foetal cells could be harvested and the normal gene added, the disease would fail to develop both in the child and *in all of its offspring*. Once the principle had been established using repair of this and other singly inherited defects, it was felt that the next stage would be to confer resistance to conditions such as heart diseases, cancer, senile dementia and maybe even doing something about getting old! Doubtless still some way off, the technology to manipulate many genes simultaneously is already with us.

Chapter 3 made the point that the incredibly long DNA molecules storing all the genes required for making a human being are wrapped up in our cells into chromosomes. We normally have 46 of them, each one with specialised sequences to ensure that the chromosome is properly replicated every time a cell divides. In 1997, a group of scientists succeeded in constructing an artificial chromosome in cultured human cells. They will now be able to assemble complex genetic programmes consisting of multiple genes, each regulated by control switches primed for activation in certain tissues or when a certain drug is taken. Once perfected, artificial chromosomes will allow complex genetic systems, perhaps even including intelligence, to be manipulated should we so desire. There are more ethical problems on the horizon.

In the meantime, nuclear transfer technology remains restricted to animals. Artificial chromosomes offer many opportunities in this area, too, but none is more exciting than the possibility of re-engineering existing animals to resurrect related species that became extinct hundreds, thousands or even millions of years ago.

Resurrecting extinct species

A species (plant or animal) is defined as a group if individuals are able to mate and produce fertile offspring. Related but distinct species are often recognisable because of a similar architecture. A lion and a domestic cat are physically similar, yet distinct species – they cannot be crossed to produce a 'lat' or a 'cion'! A donkey and horse *can* be mated to bring forth a mule but, as mules are sterile, the progenitors are regarded as separate species.

'Master genes' which make sure that the basic structure of progeny is identical with that of the parents strictly control the processes underlying the formation of a body plan. The pattern of development from a fertilised egg is similar regardless of the species; what differs is the detail. In each case the fertilised egg divides into a ball of similar cells but soon they start to look and act differently from one another. Special proteins are produced by some cells and sent to others to 'tell' them which type of cell to become, signals which activate or switch off various genes in the target cells. The extremely subtle influences of these 'master proteins' ensure (in the case of humans, for example) that the cells divide and assemble properly to make the right number of foetal legs and arms in the correct place, that each hand has four fingers and a thumb and that these appendages are of equal length on both sides. A similar set of signals causes the ribs to form in pairs at the correct sites along the backbone and that the inner organs fit perfectly inside the body cavity. In the case of four-legged animals, the body architecture is different in that the forelimbs are legs rather than hands, but the basic principle is the same for all multi-cellular living things.

To achieve this end result, the genetic information in the fertilised egg is released as a cascades of instructions as cells divide, differentiate and associate in a spatial and temporal order to produce a complex multi-cellular co-ordinated organism. This elaborate interaction of hundreds of genes and their products makes this a critically important part of development and a system which can very easily be thrown into disarray. The drug *Thalidomide* was originally developed to prevent morning sickness in pregnant women. Somehow it interfered with the flow of information normally leading to arm and leg formation so that hundreds of babies were born with limb abnormalities.

Gene mutations can have similar effects. In less enlightened times, some individuals exhibited as circus freaks suffered from this type of mutation. One of them was the 'lobster man' who had claws where his hands should have been and completely lacked lower limbs. A number of well documented cases also occur in experimental animals. One mutation causes some fruit flies to grow legs where they should have antennae. Another gives rise to a second pair of wings. A related gene in mice causes ribs to appear on a vertebra from which they are normally absent. Suffice to say, we are gradually building up a profile of how body architecture is regulated. With enough understanding of the genetic programmes governing the size and shape of living animal species, it might just be possible to use nuclear transfer technology to resurrect extinct species – if we decide that this is a good use of resources.

The woolly mammoth has been extinct for thousands of years but it is obviously physically related to the elephant. Using DNA samples from recently discovered deep-frozen mammoths, it should in the not-too-distant future be possible to compare the genome sequences of these two animals, decide on which genes are different and clone the mammoth genes into the nucleus of elephant cells growing in tissue culture. Nuclear transfer technology could then be used to inject a re-programmed elephant genome into an

1. IDENTIFY THE GENE DIFFERENCES USING A COMPUTER

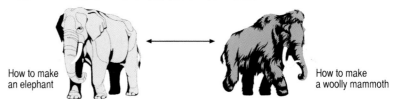

How to make
an elephant

How to make
a woolly mammoth

2. CHANGE THE GENETIC INFORMATION IN ELEPHANT CELLS

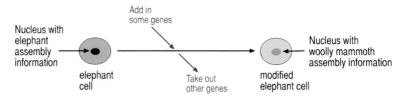

Add in
some genes

Nucleus with
elephant
assembly
information

elephant
cell

Take out
other genes

Nucleus with
woolly mammoth
assembly information

modified
elephant cell

3. REMOVE THE NUCLEUS FROM AN ELEPHANT EGG
AND REPLACE WITH THE 'WOOLLY MAMMOTH' NUCLEUS

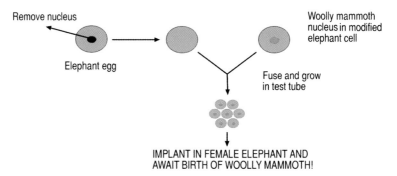

Remove nucleus

Elephant egg

Woolly mammoth
nucleus in modified
elephant cell

Fuse and grow
in test tube

IMPLANT IN FEMALE ELEPHANT AND
AWAIT BIRTH OF WOOLLY MAMMOTH!

Figure 10.4 It may be possible to 'reprogram' the cells of existing animals and then
use nuclear transplantation to regenerate extinct species.

enucleated elephant egg, grow it into an embryo in the laboratory and insert
it into a surrogate elephant mother (Figure 10.4). Once a number of mammoths of each sex were available they could be bred to produce new offspring. Despite the many technical and ecological problems that such
research would have to overcome, it is not unreasonable to expect at least
some extinct animal species to be resurrected in this fashion and the first
steps have already been taken. A 5-year plan has been proposed to recover
genetic material from stuffed specimens of the *Huia*, an extinct bird once

flourishing in New Zealand. Nuclear transfer technology will then be used in magpie cells to create an embryo and implant it into a surrogate magpie mother.

When the film *Jurassic Park* was released, no serious scientist believed that it would ever be possible to resurrect extinct species of dinosaurs. Today one cannot be so sure. The direct forebears of the *Tuatara*, a huge lizard-like creature still living today in New Zealand, were alive and well when dinosaurs roamed the earth. It may not be impossible to conceive of accumulating sufficient data on how to construct an animal so that some dinosaur species might be resurrected by tweaking the genome of their surviving cousin. As so often the case with genetics, fact can indeed be stranger than fiction.

11

All is changed – changed utterly

The evidence is all around. Fifty years ago we did not know what a gene was. Today, barely a day passes without a radio, TV or newspaper article on genetically modified (GM) food, animal clones, DNA finger printing, transgenic animals, human gene therapy or cancer chips – not to mention the ethical dilemmas arising from the latest breakthrough in gene manipulation technology. How did the elucidation of the structure of DNA put mankind into such a powerful position in so short a time? How can we control the technology, and where will it take us in another 50 years?

A revolution in biology

For hundreds of years, biologists have attacked their subject from the top down. Attempts to observe, characterise and catalogue many millions of different species of plants, animals (over 750,000 insects alone) and microbes is nothing short of a totally impossible task. Attempting to understand individual organisms is no less demanding. Multi-cellular organisms have myriad specialisations and interconnections – where does one start? Single-celled organisms like bacteria or yeast are less complex but, even with them, each individual cell contains many thousands of interacting chemical components. Generations of researchers have used reductionism with a great deal of success but reductionism means that complex questions have to be separated into smaller constituent parts, each of which must undergo detailed study before the answers can be reassembled into the bigger, more complex picture. Up to now, this so-called *empirical science* has been responsible for all the major breakthroughs in biology and genetics.

The discovery of DNA, followed by 10–15 years of further experimentation, turned the entire process on its head. It became obvious that the complexity of the living world arose directly from differences in the cells making up the enormous variety of organisms. Such differences were directly attributable to differences in proteins, in turn governed by the order of bases found in the cellular DNA – an incredibly long molecule no doubt but also, from a chemical standpoint, an incredibly simple one considering its significance.

Understand the sequence of bases in the DNA and you could understand the organism; control the sequence of DNA bases in any organism and you could control the organism.

The new concepts might have been no more than esoteric points for biologists to discuss among themselves were it not that nature likes to shuffle genetic information around in a reasonably haphazard fashion. Recombinant DNA technology hit the ground running simply because nature provided plasmids, viruses and enzymes able to manipulate DNA. People rapidly learned to adapt these processes deliberately and purposefully to move genetic information between living organisms.

The simplicity of the DNA structure added an even greater impetus to the genetic revolution because it allowed rapid cross-fertilisation of ideas and discoveries from one organism or area of genetics to others. For example, the mechanism of gene regulation was first elucidated in bacteria, yet the same overall principles can be found in plants and animals. Once the genetic code was deciphered, it was found to apply equally to all organisms. Moreover, many methods developed in one system could often be applied to others. PCR is one such technique: it was invented originally as a solution to a problem in human genetics but rapidly became the method of choice for cloning all sorts of sequences from all types of organisms.

The underlying unity provided by DNA lies at the heart of the genetics revolution. But its ability to encode information, itself relatively easy to access, has even changed the type of science used in this field. In less than 50 years, biological research has developed from manipulating cells to manipulating the molecules (genes) that make them what they are and on to using computers to manipulate the information content of those molecules. This has the effect of replacing empirical experimentation with a rational approach to problem solving. The 'new' genetics uses computer algorithms to process the enormous amounts of information generated by multiple DNA sequence analysis. This has the effect of decreasing the number of traditional 'wet' experiments (involving test-tubes, pipettes and all that) to increase greatly the amount of knowledge gained per scientist hour. Biological understanding is increasing exponentially by looking at organisms from the bottom up and its future potential is awesome. It is already starting to change the food we eat, the health systems we use and it has the potential to change the genes that we choose to pass on to future generations.

Reloading the agricultural dice
Before the 1980s, long-term breeding programmes, intensive use of fertilisers to promote growth and chemicals to control diseases, coupled with resignation at the inevitability of poor harvests at regular intervals due to

climatic/environmental conditions beyond our control, was the reality of farming even in much of the industrial world. Variable quality, quantity and nutritional values were the lot of the consumer.

For the first time, recombinant DNA technology now allows us to change all that by producing rationally designed plants. We can now adapt crops to the environment (rather than the other way round), tailor the produce to the consumer (rather than simply offer pot luck) and promote the health of the nation by increasing the number and type of nutritional compounds in the food we grow. Such increased quality and yield were quite simply impossible using traditional agronomic practices.

The lottery of *traditional* plant breeding is no different from that of animal breeding – sometimes you get lucky, mostly however you do not. It can take decades to produce a successful new plant variety and frequently even new varieties have less than optimal growth characteristics. Plant breeders are restricted to using the genes present in plants able to fertilise one another; even if they can recognise a desirable trait in another plant species, they are unable to use it.

Recombinant DNA technology can change this by removing much of the uncertainty about the outcome (Figure 11.1). Procedures have been developed over the past fifteen years to allow all the major food crops to be genetically modified. Put simply, the technology allows genes from any organism to be transferred into plant cells from which can be grown complete plants carrying the new gene. These are called *transgenic plants* because they carry the inserted gene in their reproductive cells and so will pass on the new trait to their offspring. It takes the guesswork out of plant breeding. Scientists know exactly which gene is being inserted and how its encoded protein works. Desirable traits can be introduced within months rather than taking a lifetime of effort.

Genes have been cloned from a surprisingly wide variety of sources, allowing the transgenic plants to make specific products to enhance their ability to survive pests, weeds, adverse weather conditions or whatever. A gene for a bacterial protein toxic to insects (but not to us) has been transferred into crops making them poisonous and thus resistant to attack by those insects without any adverse effects on humans. Another gene encodes an enzyme which makes the host plant resistant to certain herbicides, so allowing the use of weed-killer in fields of crops without having to resort to hand weeding. A gene encoding an anti-freeze protein in Arctic fish has been used to confer frost tolerance on plants. All of these modifications are aimed at increasing the *quantity* of produce.

Other genetic modifications seek to enhance *quality*. A genetic trick which prevents the production of the enzyme responsible for the softening

Figure 11.1 Orchards of the future: Advances in plant tissue culture permit the production and testing of trees in the laboratory without waiting for years of breeding to produce new varities. Reproduced courtesy of the United States Department of Agriculture.

process associated with ripening fruit and vegetables means that farm produce has a longer shelf life and so less waste. Seed storage proteins have been engineered to contain much higher levels of the essential sulphur-containing amino-acids. These two amino-acids are deficient in many forms of plant protein, leading to nutritional disease in poorer populations largely dependent on them and unable to afford meat or fish (Figure 11.2). Oil seed rape (canola) is currently being genetically manipulated to produce new ranges of useful and essential oils. Other projects are directed to producing food products containing higher levels of naturally occurring vitamins and other health-promoting molecules.

The phrase 'genetically engineered plant' might conjure up nightmare scenarios of Frankenstein foods, but nothing could be further from the truth. The addition of one or two *known* genes to a genome – which already has tens of thousands of them – is a much more exact science than the traditional practice of cross-breeding distantly related members of the same species thereby mixing the genetic information of two entire genomes. It is widely agreed that there are a number of conceivable problems associated with genetically engineered crops but, as each gene is unique and its gene product well characterised, such problems are *recognisable and specific* to plants carrying that particular gene. Any potential danger of gene

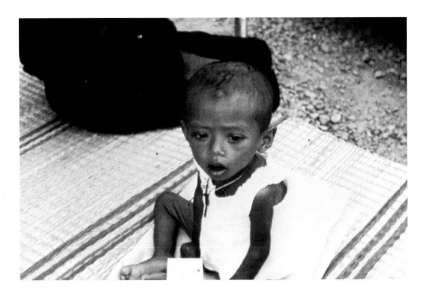

Figure 11.2 Many millions of children suffer from malnutrition throughout the third world.

manipulation lies not in the fact *that* genes are added but *which* genes are added. The main types of anxiety include possible human health problems and potential ecological impacts.

Worries about human health problems focus on the use of antibiotic genes in the plasmids used to generate the GM crops, and on the possibility that the transgenic plant will cause an allergic or toxic response in the consumer.

Antibiotic resistance
It is indeed possible that an antibiotic-resistant gene could occasionally get transferred from a GM plant to the bacteria in the gut of an animal eating the plant product. However, bacterial cells that obtain copies of such genes are unlikely to flourish for any length of time *unless that particular antibiotic is present in the bloodstream of the animal.* This is because these bacteria carry an extra DNA molecule that has to be replicated every time the cell divides, which actually slows these cells down slightly, so that over time they are out-grown by the billions of other (non-resistant) bacteria in the same gut. On the other hand, if the antibiotic is in the animal's blood stream, the resistant cells will survive in the gut whereas the other (non-resistant) bacteria will die (please see Figure 5.3 for how an antibiotic can 'select' for antibiotic resistant bacteria in a test-tube). In other words it is the selective force (the antibiotic), not the presence of genetic information (the antibiotic gene), that determines whether antibiotic resistance bacteria can flourish, reproduce and cause problems for farm animals and humans alike. Indeed it is for this reason that the rise of antibiotic resistant bacteria is directly attributable to the use of antibiotics by humans. The first observation of multiply-resistant bacteria was made during the 1950s in Japan. The *source* of the genetic information was the naturally occurring plasmids described in Chapter 5 and the DNA encod-ing this information was passed between cells by the naturally occurring con-jugation system described in Figure 5.1. But the most crucial factor – the *driving force* for the accumulation of antibiotic resistant bugs throughout the Western World has been, and is, the use (and abuse) of antibiotics in human and animal health. The most effective way to control the further accumula-tion of antibiotic resistant cells in our animal herds and in our hospitals, therefore, is by lowering the indiscriminate use of antibiotics. Antibiotic resis-tant microbes are already ubiquitous; that the same genetic information could occasionally be passed from an ingested transgenic plant to gut bac-teria cannot therefore be a reason to ban plant genetic engineering.

Allergens and toxins
Anyone at any time can become allergic to almost anything – I know a lady who developed an allergy to gold after her second pregnancy. It is therefore

entirely possible that certain individuals could suffer an allergic response arising from an encounter with a transgenic plant. But this possibility has nothing to do with gene manipulation per se. In fact, given that GM food products are subject to far more stringent regulation than new foods and ingredients produced by traditional breeding, untoward side effects are, if anything, less likely to occur in the former than in the latter. Genetic engineering allows scientists to take a *known piece of DNA* that encodes *a known protein* that performs *a known function* and to insert it into a plant genome. Probe technology allows the scientists to then find out where it is in that new genetic landscape and to characterise what effect the new DNA has on the plant cells. This type of precision is simply not available to traditional plant breeders who generate new varieties by cross-breeding existing ones, and in so doing mix tens of thousands of genes, many of unknown function, to generate a new type of plant.

Does the precision that is so characteristic of plant genetic engineering mean that a new toxin or allergen can never arise in a genetically engineered plant? – No. However, allergies and toxins are no more likely to occur in genetically modified plants than in any other plant variety. Peanuts are known to cause fatal allergic responses, and scores of other foodstuffs have severe, if less fatal effects, on millions of individuals, but these foodstuffs are readily available. All of the scientific evidence to date suggests that genetically modified organisms are no more likely to have adverse allergenic or toxic effects than any other crop. Therefore, provided that such potential side effects are properly evaluated and minimised, it is difficult to see why the use of such crops should pose any threat to the human food chain.

Ecological problems

Potential ecological concerns include fears that the new genetic information in the genetically engineered crops will escape out into the wild and/or that the genetically modified crops will upset the delicate ecosystems that exist in and around fields where they are growing.

The multi-factorial nature of ecosystems makes potential ecological impacts difficult to predict. However because the genetic information, and the nature of the gene products in question, are well understood, ecologists can use the probe technology described in Chapter 5 to design searching experiments to assess environmental effects. Such data will allow an assessment of the likelihood: that pollen from a GM crop could be transferred to non-GM crops; that pests will acquire resistance to the specific gene product being used; and that these crops will have a knock-on effect on the ecological network in the field. But, in all of this, it must not be forgotten that any negative effects arising from GM crops (and inevitably there will be some) must be balanced against those that would arise through the use of traditional insecticide and herbicide treatments (of which there are already far too many).

It is therefore critically important that the widespread planting of GM crops is undertaken only after rigorous testing and evaluation. Problems that show up are not necessarily reasons to stop the technology, but rather to alter how it should be implemented. The power and versatility of this technology is already being evidenced in the genetic answers it is providing to counteract the various shortcomings so far identified. Technologies are being developed to prevent pollen cross-contaminating other crops, while systems have been designed to avoid the use of antibiotic resistance genes when constructing these plants.

Furthermore, while current attempts at improving the quality and yield of crops may seem to have a very limited utility, it is important to realise that these are just the forerunners of much more ambitious and significant long-term research projects. Drought and salt resistance, low fertiliser requirements and more effective use of available sunlight are broad objectives being pursued by large numbers of plant biologists in many centres throughout the world. Recombinant DNA technology has initiated a major shift in agricultural practices and public awareness. It is incumbent upon us all to ensure that the prizes that accrue from reloading the agricultural dice are distributed in such a way that the poorer nations benefit as much as the rich ones. We need to implement change together if we are to meet the challenge of feeding the extra two billion people expected by the year 2020. Not only will those people be extra mouths to feed but they will need living space; if past experience is anything to go by, they will occupy prime farming land thereby exacerbating the problem and making an improvement in agricultural efficiency even more pressing.

On 'naturalness' and species boundaries

It has been argued that GM foods are 'not natural'. It does, of course, depend on what one means by 'natural' – 'able to occur in nature free from human intervention' is a reasonable interpretation. However on that basis, hardly any of the plants or animals that we eat are natural, but are all the results of 10,000 years of accidental discoveries, and deliberate breeding endeavours, by humans.

It is true to say that genetic engineering allows scientists to breach the so-called 'species barrier' and insert genetic information into organisms that would not normally have been able to acquire that information. However genome analysis reveals that genes do not exist as sacrosanct entities uniquely 'locked' into individual species. Therefore the compatibility of genetic information cannot necessarily be predicted from where it originates (we have seen in Chapter 6 that much genetic information is remarkably conserved between organisms as diverse as human and bacteria). Moreover it is notoriously difficult to define a species boundary: the Saint Bernard and Pekinese both belong to the dog species but one doesn't have to be a geneticist to see that their physical characteristics essentially preclude them from interbreeding to produce fertile offspring. Conversely, grasses, which exist as hundreds of different species, have regularly cross-fertilised one another over the aeons and in doing so produced totally new species of naturally (there's that word again!) occurring fertile hybrids. Yet intriguingly had it not been for the intervention of humans 10,000 years ago, one such hybrid would not be feeding us today: wheat originated as a hybrid of no fewer than three different grass species. However wheat seed-heads consist of closely packed large seeds and this makes the seeds very resistant to wind dispersal. Therefore, if the interspecies wheat hybrid had been left to its own devices when it first appeared, it would have been outgrown by other grasses with lighter seeds growing in the same location. However humans recognised the food value of these big seeds and cultivated wheat, thereby hand-dispersing the seeds as they sowed the crop for the next year's harvest. Thus the wheat that we still plant and eat today arose as the direct result of two separate breaches of the grass 'species barrier' followed by human intervention to ensure its survival 10,000 years ago.

Perhaps some of the fish in the deep oceans and mushrooms in the forests have escaped our influence. But don't count on it – and beware of the mushrooms: there is a very high incidence of poisoning from eating the wrong sort!

Prevention better than cure
At this time, most conditions presented in a general medical practitioner's surgery are not life-threatening and respond to symptomatic treatment. Some of the most commonly prescribed drugs are painkillers, antidepressants and sleeping tablets. By the time more serious symptoms appear (heart problems, cancer etc.), some damage at least has already been caused and frequently the health of the patient is already seriously compromised. Recombinant DNA and its related technologies will change this pattern. For an ever-increasing number of conditions, resources can now be targeted at pre-symptomatic diagnosis for susceptible persons, aiming to prevent the disease in the first place; thus, colorectal cancer is completely curable if found in time. A genetic screen for a predisposition to this disease would pinpoint individuals who should have colonoscopy performed at the age at which it normally manifests. Likewise, those individuals whose genes put them at risk

of developing artherosclerosis, can minimise the risk by adopting appropriate lifestyles.

Vaccines are probably the single most effective measure for preventing disease; recombinant DNA technology is already revolutionising vaccine design and production. The presently available hepatitis B vaccine is a recombinant protein and many others will follow. If and when an AIDS vaccine is developed, it will almost certainly be a recombinant DNA product.

Not only will these new technologies place greater emphasis on preventative medicine but the site of diagnosis is shifting from central laboratories to the doctor's surgery as ever-increasing numbers of self-diagnostic kits and services are coming onto the market. Genetic tests will follow where non-recombinant DNA products like pregnancy detection and fertility testing kits have already led. Companies are starting to set up services to carry out screens for private individuals for a number of genetically determined conditions. The molecular revolution is thus empowering individuals with extremely important health information without recourse to medical opinion. Pregnancy and fertility tests are non-controversial but access to detailed genetic information is an altogether different question (Chapter 7). The medical profession has spent many years acquiring and refining the skills to counsel individuals about genetic diseases. The results of genetic tests can have serious consequences, not only for the individual who has requested the tests, but for their relatives and children (should they have any). It is therefore essential that private genetic services should be regulated in a fashion that minimises the impact of such information.

Recombinant DNA technology will also revolutionise the discovery of new drugs. In the past, teams of laboratory workers tested thousands of different compounds on whole cells or animals, a lengthy business because it used labour-intensive manual screening procedures. It was also non-specific because compounds can have effects on a cell for a variety of reasons; it took lots of time and effort to work out which cellular component was being affected. Recombinant DNA technology has turned this process on its head because it enables scientists to begin the process by identifying a precise, well-defined cellular target, usually a protein. The gene for the particular protein is cloned facilitating the preparation of large quantities of the target protein, which can be used to screen compounds and identify compounds interacting with it. Unlike the complex screens used in traditional screening procedures, the cloned target permits simple assays executed by robots. Drug discovery has become more precise, much faster and capable of screening many millions rather than thousands of compounds (Figure 11.3).

Billions of pounds, dollars, marks and yen have been spent over the

Figure 11.3 The initial stages in drug discovery are now performed by robots. Reproduced courtesy of Shane O'Brien, Robocon.

last few years as the large pharmaceutical companies embrace this new technology. The new drugs will not only be powerful and precisely directed to specific cellular targets – they will also be tailored to individual needs. Drugs affect different individuals in different ways: some patients suffer side effects while others do not. These differences are almost always a reflection of genetic differences but, at present, the only way of testing the effectiveness of a drug is to try it with real patients. In the not-too-distant future, the doctor will use microchip technology to scan each patient's genome to identify the genes that influence the effectiveness of a particular drug. A rapid readout will show in detail how the drug will affect each individual patient. Pharmaceutical companies are already gearing up for this by producing a number of distinct versions of the same drug aimed at treating different subsections of the population according to their genetic profile.

Recombinant DNA technology has thus facilitated a gigantic step in the direction of preventative medicine, empowered individuals with self-diagnosis and totally revolutionised the discovery and use of pharmaceutical drugs.

In pursuit of perfection

One of the most ethically contentious areas of the genetic revolution is the use of information acquired during prenatal screening programmes. At the present time, many countries permit the termination of pregnancy if the foetus is going to develop and suffer from a genetic condition seriously compromising its future quality of life. Before long, chip-based technologies will be able to screen for any number of genes at one time, so offering unprecedented precision in characterising the genetic make-up of the unborn. Eventually it will be able, with uncanny accuracy, to predict traits such as eye and hair colour, IQ, height and general disposition. Will such information lead to requests for termination that have little to do with preventative medicine and more to do with obtaining preferred traits in our offspring?

On the other hand, if we succeed in developing germ-line gene therapy to repair defective genes in the as yet unborn, who will decide what constitutes 'defective' let alone 'desirable' genes? Few people would argue with repairing defective genes causing cystic fibrosis, inherited cancers, etc., but very soon we run into trouble with a definition for defective. The gene for sickle cell anaemia causes a severe dysfunction when the individual carries two defective genes. If an individual has one normal and one defective gene, he will have sickle cell trait and suffer occasional bouts of anaemia. But he will also be resistant to malaria which people carrying two completely 'normal' genes are not (Chapter 2). So is the sickle cell gene defective or desirable, an advantage or a disadvantage?

What about mental illness? If we were to eliminate manic-depression from the population, would we also lose individuals with incredible creative energy? What about height? In a population of tall individuals, 5 feet 10 inches (1.78 metres) for a male is short; should it be engineered up? What about IQ? Would you choose to tweak IQ genes in your egg or sperm to allow your offspring to be more intelligent? Then there is the matter of homosexuality: in our fairly enlightened society, being gay is becoming increasingly acceptable yet, even now, for many it means a life of quiet desperation. Suppose that a definite genetic link to homosexuality were found: would homosexuals wish it on their offspring should they have any? Would heterosexual couples seek to avoid having children with that gene combination?

Germ-line gene therapy will almost certainly become available; for some neurological disorders such as Tay Sachs disease and Huntington's chorea, it may be the only way to alleviate the condition. Can we really deny it to the families of such individuals? If allowed, germ-line gene therapy could genetically repair many medically important conditions, minimise the number of patients seeking somatic gene therapy and theoretically produce a population of individuals devoid of a negative genetic legacy. Yet 'negative' is a relative term; we have insufficient knowledge to assess the usefulness or not of many apparent mutations. Our definition of dysfunction is very subjective and history is littered with the corpses of millions of unfortunate individuals whose genes did not fit into the then prevailing political framework. Despite their many drawbacks, humans face these moral dilemmas at the start of the twenty-first century because we are great survivors and innovators. A population is greater than the sum of its individual parts and, the greater its diversity, the greater its flexibility. Humans have gone from swinging stone-age clubs to swinging golf clubs 250,000 miles above the earth, not despite, but because of the combination of genes ('good' and 'bad') that we collectively possess. Let us tread carefully, very carefully, lest this power seduce us into creating a population devoid of individuality – clones of nothing greater than the lowest genetic common denominator.

Epilogue

The scientist and the saint?

Jacques Monod, together with François Jacob and André Lwoff, was awarded the Nobel prize in 1965 for uncovering the secrets of gene regulation. Throughout his life he maintained that humans exist in an unfeeling universe from which we emerged by chance. In October 1971, this giant amongst scientists appeared on a Canadian TV programme where, defending this thesis, he argued that the future destiny of the human race was inexorably locked up in its genes. All the while a middle-aged lady appearing on the same programme sat quietly with her head bowed. When pressed by the compère for her views, this fragile individual, who was herself to receive a Nobel Prize some years later, simply lifted her head and remarked: 'I believe in love and compassion.' The Nobel Prize Mother Teresa received was not for ingenious experiments executed in a laboratory but in recognition of her services to humanity.

While there will be many interpretations of her remark, we do need to think through the consequences of all our actions to try to secure the greatest benefit for the most people while wreaking the least amount of harm.

The choice is ours

As we have seen, it is scientists who are providing new knowledge about genetics – but it is society as a whole which must decide what to do with it. Not surprisingly, different societies – and communities within societies – have very divergent expectations for what genetic engineers should and should not be allowed to do. Controversy surrounds any significant technological advance and genetic engineering is a hotbed of such argument. It is vitally important that a technology so powerful is implemented with due deference to opinion from all sections of society.

What is 'written' in an individual's genes is intensely personal yet it can have an impact on both the family and society at large. The sensible use of genetic knowledge could help to prevent the exclusion of individuals with a genetic predisposition to certain illness from jobs and insurance cover.

Similarly, it will ensure that the powerful new chip technology, with its unprecedented precision, is used wisely in prenatal diagnosis. We need to think very carefully about including non-medically relevant genes in such chips, and whether and how insurance companies or over-stretched health services might try to prevent the birth of individuals with genetic defects – and whether they would be right or wrong to do so.

The release of genetically engineered organisms into the environment is, at the time of writing, a topic of intense public discussion. What 'the people' actually think is by no means clear but self-proclaimed 'environmental protection' pressure groups, supported by sections of the press and broadcasting, exert constant political pressure against the production and use of GM corps. Emotions are heightened, 'activists' sabotage experimental growth plots and balanced debate becomes very difficult. The public, battered by partisan opinion from both sides of the argument, don't know what to make of it: are GM crops the saviours of humanity (as their commercial protagonists would have us believe) or tantamount to the work of the devil (judging by the way many of the environmentalist talk). Can we ever know the exact consequences of introducing genetically modified plants into the field? How much do we need to test? How relevant to other parts of the world are 10 years of US experience? How circumspect do we need to be before pressing on?

Is there a real threat from the ever-increasing power of multinational companies in their pursuit of market share? It can be argued that holding patents of plants and genetic engineering processes is an acceptable level of self-interest to ensure the research efforts are justly repaid. Are the seed companies *forcing* unwilling farmers to use their products (an environmentalist claim) or are they offering agricultural stability, more stable incomes and better consumer satisfaction (the commercial argument)? One cannot expect altruistic behaviour from biotechnology companies any more than from other commercial establishments (as mentioned in Chapter 9, companies are not charities but exist to make profits for their shareholders).

However, most people would probably agree that recombinant DNA technology should be used, where possible, to help close the gap between the richest and poorest countries. A fifth of the world's population already owns 80% of its wealth; how can we redress the balance? If large scale famine is to be avoided, food production must increase by 40–50% by the year 2020 while using a smaller land area for cultivation. Given that most of this additional food will be needed in developing countries, should it be produced there or in developed countries, where agricultural efficiency is so much greater? How can we best introduce the benefits of genetic technology to the poorer countries with both limited resources and limited traditions for using all new forms of technology? International agencies are already struggling to deal with these problems and many see the transfer of the genetic technologies from the industrial regions as a key to this drama. It will happen only if the political will (ultimately the will of the public at large) is there and will be effective only if the recipients are properly trained.

A lack of appropriate healthcare combines with starvation to limit the average life expectancy in sub-Saharan African countries to 50 years, versus 75 in western countries. Unlike agriculture, which could gain from recombinant DNA technology to increase crop productivity, enhanced healthcare would benefit most by simple improvements in the provision of clean water, education and vaccination: every incremental expenditure of money might be best spent on improving the drains. The problem of economics is very important here. Biotechnology is not cheap. While fundamental research, paid for by governments and carried out in universities and public sector institutes, seeks to understand the nature of disease, drugs and treatments (both very expensive) are developed by commercial companies. They are so expensive that inevitably they are directed to people (or their insurance companies or health services) who can afford to pay. There is accordingly a big problem about who pay for drug developments to deal with widespread and very debilitating illnesses affecting the poorer parts of the world. That is a major moral issue conveniently ignored by most people when they go out to buy a new car or sign up for their second overseas vaction of the year.

Future historians may view the genetic revolution as the one of the most defining events in the history of mankind. In the near future we will have acquired sufficient knowledge, understanding and power, rationally to alter our own and any other species we so choose. We should eventually be able to eliminate many – perhaps most – forms of disease, extend life expectancy and dominate, even more than we do now, the enviroment in which we live. Are we going to use our knowledge of genetics and all of the other new marvellous technologies to make things fairer and better? Or make a mess of it? No doubt, as usual, it will be somewhere in between. It's up to you and me.

Index